HIGH PRAISE FOR
FRONTIERS II

"Touches upon many different fields of science, including the discovery of whale fossils with feet, prospects for making Mars livable for humans, and the benefits of making robots look less like people."
—Science News

"Entertaining and informative . . . touches on an astonishing variety of topics. . . . It is no surprise that Asimov spoke knowledgeably and engagingly on so many topics." *—Bookpage*

"Offers a solid overview of a wide variety of scientific subjects and current issues."
—Charleston Post and Courier

"Lively . . . Asimov fans will appreciate this collection of the science polymath's last pieces."
—Publishers Weekly

"Offers dozens of essays on the latest scientific discoveries in such fields as genetics, space colonies and dinosaurs." *—Associated Press*

ISAAC ASIMOV was the author of many major non-fiction science books, including *The Genetic Code*; the two-volume *The Human Body* and *The Human Brain*; the three-volume *Understanding Physics*; *The Exploding Suns*; and the recent *Atom: Journey Across the Subatomic Cosmos*. The year 1990 marked the fortieth anniversary of Isaac Asimov's first of over 475 books, collections, and anthologies. He died in April 1992 at 72. JANET ASIMOV, his wife and close collaborator in recent years, has contributed one-quarter of the contents of this, his final book.

FRONTIERS II

More Recent Discoveries About
Life, Earth, Space, and
the Universe

Isaac and
Janet Asimov

TRUMAN TALLEY BOOKS / PLUME
NEW YORK

TRUMAN TALLEY BOOKS/PLUME
Published by the Penguin Group
Penguin Books USA Inc., 375 Hudson Street, New York, New York 10014, U.S.A.
Penguin Books Ltd, 27 Wrights Lane, London W8 5TZ, England
Penguin Books Australia Ltd, Ringwood, Victoria, Australia
Penguin Books Canada Ltd, 10 Alcorn Avenue, Toronto, Ontario, Canada M4V 3B2
Penguin Books (N.Z.) Ltd, 182–190 Wairau Road, Auckland 10, New Zealand

Penguin Books Ltd, Registered Offices: Harmondsworth, Middlesex, England

Published by Truman Talley Books/Plume, an imprint of Dutton Signet,
a division of Penguin Books USA Inc. Previously published in a
Truman Talley Books/ Dutton edition.

First Plume Printing, August, 1994
10 9 8 7 6 5 4 3 2 1

All of the articles in this book originally were written for the Los Angeles Times
Syndicate.

Ⓟ REGISTERED TRADEMARK—MARCA REGISTRADA

LIBRARY OF CONGRESS CATALOGING-IN-PUBLICATION DATA
Asimov, Isaac, 1920–1992
    Frontiers II : more recent discoveries about life, earth, space,
  and the universe / Isaac and Janet Asimov.
        p.   cm.
    ISBN 0-452-27229-7
    1. Science—Popular works.   2. Technology—Popular works.
  3. Astronomy—Popular works.   I. Asimov, Janet.   II. Title.
  [Q126.A73   1994]
  500—dc20
                                        93–48686
                                          CIP

Printed in the United States of America

To the memory of
my beloved husband,
Isaac Asimov

CONTENTS

Contents

TABLE OF ATTRIBUTIONS

Three-quarters of the articles were written by Isaac Asimov.

Written by Isaac and Janet Asimov:

The Power of Proteins • Vital Cooperation • Left,
Right • Genes in Action • The Beautiful Microbes
Delightful Diversity • Noise • The Park Phenomenon
Music, Always • The Usefully Small • Galaxy Update

Written by Janet Asimov:

Bony Heritage • Dinowalk • Gone—Again and Again
Heady Stuff • Mantle and Core • Water—the Circulation
Below • Air—the Circulation Above • More on Venus
Mars for Humans • More on Comets • Our Own Private
Sun • Out of the Sun • More on Meteors • Garbage!
The Critters Within • More Replication • Nanomagic
Fantastic Fullerenes • Plant Power • Plant Help
Cockroaches and Computers • Once and Future Robots
Star Clusters • The Black Hole Tango • Far-Out Reality

Introduction

An interest in science has many rewards, but the one I appreciate the most is the exciting sensation of being on the frontier. I came from pioneer ancestry, to whom the American frontier was real and vital. That frontier has disappeared, but there will always be frontiers in all aspects of science, for solving a scientific problem opens up horizons containing new problems to exercise human curiosity and thought.

My husband, Isaac Asimov, loved science and loved writing about it. His weekly science columns for the Los Angeles Times Syndicate were collected in *Frontiers*. *Frontiers II* contains the rest of Isaac's columns and some of mine. I began writing them when Isaac was ill in the winter of 1991–92, and continued after his death in April 1992.

Despite careful predictions, the future is unknown—until it becomes the present. Scientific discoveries going on *now* are presented in this book in the hope of stimulating imaginations while making our complicated world a little clearer.

—*Janet Asimov*

I
LIFE: PAST, PRESENT, AND FUTURE

The Power
of Proteins

Protein is a remarkable word and an even more remarkable part of the universe. *Protein* comes from a Greek word meaning "of first importance"—and so it is, for without proteins, there would be no life.

Protein was the name suggested by that inveterate namer of organic compounds, the Swedish chemist Jöns Jakob Berzelius. The Dutch chemist Gerardus Johannes Mulder used Berzelius's suggestion in 1839 when he worked out a basic formula for what were then called "albuminous compounds," like egg white (casein) or blood globulin.

Carbohydrates and fats supply carbon, hydrogen, and oxygen (in various patterns), but proteins supply, in addition, nitrogen, sulfur, and often phosphorus. Proteins are complicated, and scientists are only now discovering the full extent of those complexities in the living cell.

Early methods of organic analysis were too crude to decipher the structure of proteins, but it was possible to analyze their building blocks of amino acids, which have a basic pattern of hydrogen and nitrogen atoms, a group of carbon, hydrogen, and oxygen atoms, and a branch group of atoms that identifies a particular amino acid.

After another Swedish chemist, Theodor Svedberg, invented the ultracentrifuge in 1923 (for which he received the Nobel Prize), scientists were able to determine the molecular weights of many proteins on the basis of their rate of sedi-

3

mentation. The results were astonishing, for some proteins turned out to have molecular weights in the millions, indicating that their structure was exceedingly complex indeed.

Newer technology came along to help examine the structure of proteins—nuclear magnetic resonance, chromatography, spectrophotometry, X-ray diffraction, and so on. It was found that in spite of the theoretically mammoth number of possible amino acids, proteins here on Earth contain only twenty varieties. It's quite probable that steak from another planet would not agree with a Terran.

For years, scientists believed that what they discovered about proteins in their test tubes was true of proteins in the living cell, but that turns out to have been a bit of hubris. Unanswered questions about cellular protein are keeping scientists busy, for it seems that proteins do not fold, spindle, or mutilate by themselves. They need help.

Folding is the key word. A protein's amino acid components have to be arranged correctly in order for the right doohickeys to be in the right place to do the right job. You can't have a nitrogen atom waggling off there when it should be here, up against something else. Mary-Jane Gething and Joseph Sambrook have described the fascinating functions of cellular proteins called "chaperones." These seem to exist in order to (1) help a complicated protein molecule get folded properly, (2) stabilize partially folded intermediates or inactive proteins, (3) rearrange cellular macromolecules being assembled and disassembled, (4) protect proteins that are under environmental stress, and (5) pick out proteins for destruction.

All of this research may sound esoteric, but it's of vital importance. You are alive—why not understand as much about life as possible? The new molecular biology research on proteins may make it possible to understand and cope with various diseases now incurable. There could be better medicines, designed to help cells heal themselves and to do no harm. Using chaperones, biotechnology might be able to produce important human proteins in quantities undreamed of now.

Proteins are also described as assembly line producers, as pumps for transfer, and as the engines that literally make life move. At a recent conference, the big question was—how do protein machines use the chemical energy? Some people think the trick is done by changing shape, but others disagree. Finding out the truth is tricky, since it's necessary to make an inventory of the parts involved, identify the chemical intermediates of each reaction, measure the rate constants for the transitions, and describe the detailed structure of the protein to understand how the various chemical reactions are made to work. Not one of these steps is sufficiently known right now.

Biochemists and molecular biologists will continue their research on proteins, so keep posted. When Freeman J. Dyson was asked which came first in the evolution of life, proteins or DNA, he answered, proteins.

Understanding those proteins will help humanity delve further into the mysteries not only of cellular pathology but of the origin of life itself.

The
Oldest Protein?

The oldest proteins may have been detected in 1991 by a group at Los Alamos National Laboratory in New Mexico, headed by W. Dale Spall. This was done through a study of bones. Bones are by no means simply dead minerals. They contain elaborate structures of proteins that exist in the bones even after an organism is dead. Naturally, the proteins slowly decay, but, under some circumstances, they don't decay entirely.

The bones that have yielded the (possibly) oldest proteins are themselves most unusual for they are taken from the longest and largest dinosaur that has yet been discovered. This is the "seismosaurus," the first part of whose name means "earthquake," because the notion is that when it walked it shook the Earth. The seismosaurus was excavated in central New Mexico, and it is about 160 feet in length—almost two city blocks long.

Naturally, a creature like the seismosaurus has huge bones, and deep in the interior of those bones, proteins might exist protected from the outside world. The Los Alamos team drilled a core out of one of the huge vertebrae of the seismosaurus, used solvents to dissolve the rocky materials of the fossil, and found material that seemed to be proteins of two, or perhaps three, different types.

If there were indeed proteins inside the seismosaurus vertebrae since the animal died, then they are perhaps 150 million years old. This is a record, for the oldest proteins that scientists have hitherto worked with were only 1 or 2 million years old. Unfortunately, there was not enough material to be able to identify the proteins. In ordinary bone, the most common protein is one called "collagen," but the material obtained from the seismosaurus is not that. If amino acids can be identified in the material, then what's there are proteins.

There are always problems, of course. For instance, in the 1960s, meteorites fell and were analyzed and were found to contain amino acids. Naturally, it was first assumed that these were signs of meteorite life.

However, there are two kinds of amino acids: L and D. The L-amino acids are present in living things, but the D-amino acids are not. If then, amino acids had been formed in the meteorites by ordinary chemical processes, there should be both L- and D-amino acids in equal quantities. This proved to be so, but it was not evidence of life.

The meteorites here in question were seen to fall and were freshly analyzed. Meteorites that have been lying in the soil for quite a time may well yield L-amino acids, not because there was any connection with life, but because the Earth is full of L-amino acids. They exist in groundwater and wherever organisms live. As a result the meteorites become contaminated.

Was the protein matter obtained from seismosaurus bones the result of such contamination? Spall maintains that since the seismosaurus bones were extraordinarily well-preserved, the protein could be uncontaminated by groundwater, although Spall himself admits this is a possibility. Stephen A. Macko, a geochemist at the University of Virginia, has helped analyze proteinaceous material from dinosaur bones, and he points out that even if amino acids are present, they may not be part of protein molecules, but of other types of substances.

So the situation remains uncertain, but it would be of great interest to paleontologists if the material *was* protein and if it *was* uncontaminated. Then, if enough of the material is accumulated, it may be possible to determine the order in which amino acids occur. This can be compared with the order in other dinosaurs, in living reptiles, in birds, and so on. Macko points out that if this can be done, it may be possible to spot relationships among different groups of animals that, at present, we cannot determine.

We can, in that way, form a new "tree of life" that might be better than those that now exist, and which may help us work out the course of evolutionary change in primordial times.

The Categories of Life

Recently, Carl Woese, a biologist at the University of Illinois, announced a new way of dividing all living things into groups and subgroups. The division is based on the structure of ribonucleic acid molecules (RNA) in tiny granules called "ribosomes" that are present in all living things without exception.

The ribosomes are the sites on which proteins are manufactured according to the nucleic acid blueprints present in all cells, and life cannot exist without proteins. The ribosomes must have been formed in the very earliest cells, and the RNA molecules making them up have maintained themselves with only minor changes for billions of years.

Some minor changes have, however, established themselves. The RNA molecules are, for instance, made up of chains of smaller molecules called "nucleotides." Every RNA molecule in the ribosomes has a hairpin curve at a certain place. In that curve there are either six or seven nucleotides. Every cell with six nucleotides belongs to one grand group of life forms. There are other small differences that divide the seven-nucleotide cells into two other grand groups, making three altogether.

In the old days, life forms were divided by appearance. When we play "twenty questions," for instance, we still divide all objects into "animal, vegetable, or mineral" because we consider all living things to be either animals or plants. There are microscopic life forms, too, which we used to assign to animals and plants. Amoebas, for instance, were considered animals, and bacteria were considered plants.

8

Eventually, as we studied cells more closely, we found that there were two major kinds of cells. In the more primitive cells, the nucleic acid molecules were spread throughout the cell. These made up the "prokaryotes" and included bacteria, some with chlorophyll (which allows them to live on the energy of sunlight) and some without. A more complex and advanced type of cell has its nucleic acids localized in a small region within the cell called the "nucleus." Cells like these are called "eukaryotes." Plants and animals are made up of cells that are eukaryotes—plants having cells with chlorophyll, animals without chlorophyll. Human beings are made up of cells that are eukaryotes and that, of course, contain no chlorophyll.

All this is overthrown, however, if, as Carl Woese does, we consider the molecular structure of the RNA in ribosomes. It seems that of the three grand divisions, two are single-celled life forms with the cells prokaryote in nature. (Prokaryote cells have never developed organisms that are made up of more than one cell.) One grand division of prokaryotes is called the "Bacteria." A second grand division of prokaryotes Woese calls the "Archaea," which resemble bacteria very strongly in appearance but have certain characteristics in their RNA molecules which show that they lean toward the eukaryotes.

Apparently the Archaea have begun to develop eukaryotic characteristics, and from them there evolved the third grand division of life, which Woese calls "Eucarya." These are made up of cells containing nuclei.

The Eucarya are the only life forms that ever developed in the multicellular direction. Some Eucarya developed into organisms made up of trillions or even hundreds of trillions of eukaryotic cells (we are ourselves examples of this). Not all have. Of the six main divisions of the Eucarya, three are exclusively one-celled in nature.

Of the remaining three, two are plantlike in nature. One, without chlorophyll, makes up the "fungi." Most of these are one-celled, but there are multicelled forms like mushrooms.

The second contain chlorophyll and are the familiar "green plants." Again, some of these are single-celled, like algae, but there are gigantic forms like sequoia trees.

Finally, the last group of the Eucarya are animals. These are, from our own standpoint, the most advanced of all and include giant whales that may weigh up to 150 tons. (Some plants are heavier still, but most of a tree is supportive woody tissue that is not truly alive, whereas animals consist almost entirely of living tissue.)

In a way, this puts us in our place. About 3.5 billion years ago, it remained single-celled and prokaryotic for over *2 billion years*. It was only about 1.4 billion years ago that the third grand division, Eucarya, developed, and for the next 600 million years, they remained single-celled life forms as well.

Only about 800 million years ago, the first multicellular eukaryotes began to appear, and at first only as different kinds of worms. Only 600 million years ago life forms as advanced as shelled sea life developed. We humans belong to the subphylum Vertebrata, which started only 500 million years ago. Not only we, but all the animals to which we are related, down to the most primitive fish, are Johnny-come-latelies on Earth.

Our Ancestor, the Coelacanth?

Three German scientists, Tomas Gorr, Traute Kleinschmidt, and Hans Fricke, have reported a kinship that makes it seem as though coelacanths are the direct ancestors of the tetrapods.

The tetrapods are "four-footed animals" and include amphibians, reptiles, birds, and mammals—and us.

The coelacanth is a fish known to exist in the days before the dinosaurs but once thought to be long since extinct. Then, in 1938, a fisherman trawling off the coast of South Africa caught a strange creature in his net. He eventually showed it to a South African ichthyologist, J. L. B. Smith, who identified it as a coelacanth. Since then, it has been found that coelacanths live in the Comoros and that fishermen bring up one now and then. The coelacanths live at depths of two hundred meters, and altogether 170 specimens have been brought to the surface. Unfortunately, they cannot withstand the loss of water pressure and only live a few hours when surfaced.

In 1986, Hans Fricke, making use of a specially designed two-man submarine, managed to get down low enough to follow the coelacanth in its home waters and found that the coelacanth could swim in every direction, including backward and upside down. Fricke and his two colleagues published a paper in May 1991, in which they showed that blood taken from coelacanths possessed chains of amino acids that strongly resembled similar chains of amino acids in tadpoles.

Amphibians are the least advanced of the tetrapods, and tadpoles are the least advanced of amphibians. They hatch out of eggs in tiny fish form, with tails and gills and without legs. Only as the tadpoles develop do the tails disappear, the gills turn to lungs, and legs form.

Consequently, if tadpoles resemble coelacanths in their amino acid chains, we can only suppose that the coelacanths have some sort of association with tadpoles. The coelacanths are probably the ancestors of tadpoles and of all the other tetrapods, including man. Although the amino-acid chain doesn't show up in the adult frog, the relationship between coelacanths and tadpoles is closer than that of tadpoles with any other creature that might possibly have been a tetrapod ancestor.

Another important point is that the coelacanth has six fins, four of which move in pairs. Fricke, when he studied the coelacanth underwater, found that these four fins moved in a way common to four-legged animals but not to fish.

As usual in these cases, the connection between coelacanths and tetrapods is by no means accepted by all biologists. The complaint is that amino-acid chains can be interpreted in different ways and are by no means proof of a connection.

In fact, there is another kind of fish, the lungfish, also long thought to have been extinct but rediscovered in the nineteenth century. No less than six living species of lungfish are now known.

Lungfish have primitive lungs and can, and do, gulp air when that is necessary. They can manage to get out of brackish water and make their way to larger pools. When this is not possible, especially in the summer, they can cake themselves in mud and stay largely immobile for the season. This is called "estivation," which is parallel to hibernation among other animals. Some lungfish are even likely to drown if they cannot be in a position to gulp air now and then.

A number of biologists believe that lungfish are the ancestors of the tetrapods and that the presence of lungs is more important than the presence of coelacanth fins or of amino-acid chains. The chances are that the argument will not settle down for a long time, but I am on the side of the coelacanth. I'm impressed by the amino-acid chain, and I'm not impressed by the primitive lung.

Consequently, I believe that the coelacanth, which existed not only before the time of the dinosaurs, but before the time of the tetrapods generally (the dinosaurs were advanced tetrapods), could be looked upon as our ancestor.

It's really amazing that the coelacanth, having served as ancestor of the tetrapods and having had descendants who developed in enormous ways, should nevertheless maintain itself in a form that did not change and did not develop. It is 400 million years old and has scarcely changed in all that time.

What a pity it cannot withstand surface conditions. We could study it so much better if it could.

Invading the Land

Recently, British paleontologists, led by Andrew J. Jeram of the Ulster Museum in Belfast, uncovered the oldest remnants of land life and reported them to be about 414 million years old. This is older than scientists had earlier thought, but it doesn't change the fact that land life is a relative newcomer to the planet.

The Earth is about 4.5 billion years old, and by 3.5 billion years ago at the latest, its waters swarmed with tiny bacteria-like forms of life. For at least 3 billion years afterward, life was confined to Earth's waters, and the land remained absolutely sterile.

It was only in the last ninth of Earth's existence that living things moved out to colonize the land.

This is not surprising. Whereas Earth's waters, especially the ocean, are a stable and friendly environment for life, dry land is dreadful. For life to venture out into the hostile surroundings of the dry land is roughly equivalent to human beings venturing into outer space. And whereas human beings are aided in their project by all the technological devices they have created, life forms invading land could only make use of the dreadfully slow changes produced by hit-and-miss biological evolution.

Compare the sea and land. In the sea, there is no such thing as weather. Conditions are stable. Temperatures don't change much, and while the surface may be roiled by storms, the regions not far beneath are quiet. On land, temperatures reach highs never experienced in the ocean, and swoop far below zero. There is wind, rain, snow, sleet, and all the manifold manifestations of a restless atmosphere.

In the water, buoyancy virtually eliminates gravitational pull so that you can have whales, weighing up to 150 tons, who are able to move about freely in three dimensions. On land, gravity is an endless pull and life forms must develop tissues (wood or bone) that will support them against that pull, or they are doomed to remain very small.

On dry land, life forms must find ways of storing water, and of using limited amounts of that water to eliminate waste, whereas in the sea, neither process is any problem. The result is that even today, after hundreds of millions of years of adaptation to land life, the Earth's land remains less rich in life than do its waters.

Of course, there are also advantages to land life. Since air is far less resistant to motion than water is, land animals don't have to be streamlined. They can develop appendages, and this reaches its climax in the human arm and hand. Then, the existence of free oxygen on land means we can have fire—something not possible in the sea—and it is with fire that human beings have built their technology. The equally brainy dolphins trapped in the sea cannot build one.

But if land was such a hostile environment, why did life forms invade it? They didn't do it because they "wanted to," I assure you. They did it because they had to. The ocean was crowded with life, on an eat-or-be-eaten basis. It was the shallow regions bordering the continents that were richest in life (and still are).

This meant that any life form that could somehow crawl up onto the beach and withstand a period of dryness at low tide was less likely to be eaten by predators, most of whom

14

had to remain in water at all times. Eventually, predators developed that could also last through low tide, so that there was a continual push to move higher up on the beach and stay water-free for longer and longer periods. Eventually, some forms of life could remain on dry land indefinitely.

The usual feeling is that the first organisms to invade the land, more or less permanently, were very primitive plants that had no roots and consisted of a simple forked stem without leaves. They made their timid appearance at the edge of the shore perhaps 450 million years ago.

It was only after plants had appeared that animal life could follow, using the plants as food. The first animals to make their way out on land seem to have been simple arthropods—spiderlike creatures. The earliest date at which they emerged on land was, until recently, given as 400 million years ago.

However, Jeram and his group worked with ancient rocks from the city of Ludlow in Shropshire, England. They treated the rocks with hydrofluoric acid, which can dissolve the rocks and leave behind tiny bits of shell fragments. These, when put together carefully, seem to represent the bodies and legs of small, primitive spiders and centipedes about one-twentieth of an inch long. Since the rocks they were found in were 414 million years old (as measured by the usual geologic techniques), so were these land creatures.

It was not till about 40 million years later, however, that the first backboned animals (primitive amphibians) appeared on land. It was these amphibians that were the ancestors of all modern amphibians, reptiles, birds, and mammals—including ourselves. Our own land history, then, goes back about 370 million years.

The Egg
on Land

In December 1989, T. R. Smithson of Cambridge Regional College reported the finding in Scotland of an ancient fossil reptile that dates back about 338 million years. This may not seem very important, since most of us don't think much of reptiles —snakes, lizards, turtles, alligators. But if we dismiss them we are wrong, for they are extremely important creatures. To see why, we must go back in time.

Some 450 million years ago, the Earth was about 4 billion years old, and life had existed on it for about 3 billion years at least. In all that time, however, life had existed exclusively in water. The dry land was sterile.

About 450 million years ago, however, the first plants began to creep out on shore and the tidal area began to turn green. The earliest land plants had no roots or leaves, but the pressure of evolution produced these, as well as stems, and by 400 million years ago, the first forests covered the land surface.

Why did it take so long for life to emerge on land? Well, land is a hostile environment, with a strong gravity pull, extremes of temperatures, and the possibility of drying out. It took hundreds of millions of years for life to develop the devices that could counter these difficulties. For almost 40 million years, plant life lived in isolation, and then animals began to follow. The plants were a plentiful food supply, and any animal that could develop ways of enduring land life could multiply freely.

The first animals that emerged on land were primitive

spiders, scorpions, snails, worms, and, eventually, insects. They were all small creatures. They had to be small so that the pull of gravity would not be strong enough to immobilize them.

For *large* land creatures to develop, they would need limbs and bodies stiffened by bones. In short, we would need vertebrates. Four hundred million years ago, there were swarms of vertebrates, but they all lived in the water. They were the fish, and, to this day, they dominate the oceans of the Earth.

Some fish had delicate fins that were fitted chiefly for steering and propulsion, but some had sturdy, fleshy fins that were almost like small legs. These fleshy-fin fish were, by and large, not as successful as ordinary fish, but they had one advantage. If they lived in a pond that grew brackish or threatened to dry out, they could stump across land into another and larger pond.

Such fish developed the ability to stay out on land for longer and longer periods. They developed primitive lungs that allowed them to breathe on land, and they were the "amphibians," which seem to have made their first appearance about 370 million years ago. They were the first land creatures, many large and formidable.

Amphibians had one important flaw, however. Their eggs had to be laid in water, and while they were developing to adulthood they remained fishlike. The most familiar amphibians today are frogs, and we know that their eggs develop into tadpoles, which only gradually become frogs. On the whole, then, amphibia were tied to the water and were not truly land creatures.

Then came the reptiles, which had a new kind of egg, one that contained a complex embryonic membrane called an "amnion." The egg had a shell that allowed air to enter and leave but not water. It came with a water supply sufficient for the developing embryo, and the wastes were deposited within the amnion. This "amniote egg" could be laid and hatched entirely on land, so the reptiles were the first true land vertebrates.

For over 250 million years, they dominated the land, producing the most magnificent creatures who ever clumped over the Earth, the animals we refer to as "dinosaurs."

It is important to remember that birds are simply modified reptiles. They are warm-blooded and have feathers, but they lay reptilian eggs with an amnion.

Mammals are also modified reptiles. They are warm-blooded and have hair, but when they first appeared, some 200 million years ago, they laid reptilian eggs with an amnion.

The birds and mammals were *not* successful while the reptiles dominated the Earth. They were small creatures which survived only because they were largely unnoticeable. They were the equivalent of sparrows and mice, peeping about in the shadows of the great reptiles. If the dinosaurs had not been wiped out (probably by an asteroidal collision) 65 million years ago, birds and mammals might still be insignificant today.

It was the development of the amniote egg that made everything else possible, including us. So when we find the oldest reptile, we may have the creature who invented the egg on land—and that's supremely important.

What Teeth Can Tell Us

The hardest part of the vertebrate body is the teeth—naturally, since they have a difficult job. This means that when ancient forms of life are fossilized, the one part that is sure to survive all the vicissitudes of geologic change is the bony head with teeth—sometimes only the teeth. Therefore it is important to glean as much information from them as possible.

Teeth are characteristic of vertebrates and began, I believe, with sharks. Teeth are found in most of the land vertebrates, but not all. Birds, for instance, do not have teeth. But they had teeth to begin with. The early extinct bird archaeopteryx had lizardlike teeth, but, with evolutionary change, it disappeared.

Birds now have beaks, which, for their diets, are far more efficient. Birds have a higher temperature than we do, and to maintain that high temperature, they must constantly be eating. Their beaks allow them to efficiently peck away at food such as small seeds and insects. Turtles and tortoises also lack teeth. They have horny beaks that are nowhere near as efficient as those of birds. However, turtles are long-lived and slow-living and don't require efficiency in that respect.

Early mammals have teeth that are very much alike—just a series of choppers. With evolutionary change, the teeth also developed and became different in size and function. We ourselves have incisors that are cutting teeth and molars that are grinding teeth. Many animals have fangs that do the ripping and tearing of food.

Teeth also develop in odd ways. Elephants and walruses have tusks, and the narwhal has a single long tusk of spiral shape. Poison snakes have fangs that conduct a poison, and their bites can be easily fatal. But all in all, human teeth may be the most useful of all.

It would be of great use to paleontologists if information could be squeezed out of teeth. It doesn't seem likely, but scientists are now busily engaged in trying to do just that.

Gregory M. Erickson of the Museum of the Rockies in Bozeman, Montana, counted the tiny lines of growth in the teeth of dinosaurs. He compared the results with similar investigations of the teeth of alligators, the closest living relatives of the dinosaurs. He maintained that every line of growth represents a day in the life of the large, extinct creatures. Naturally, this yielded information about how long it took a dinosaur's teeth to develop, and more generally still, how long the dinosaur lived.

This is not an easy test to perform. The first problem is to get the alligator teeth. The alligators are not killed for their teeth; that would be unconscionable. Instead, teeth are taken from alligators that have already been killed for their skins. (I'm bitterly against that, too.) The teeth are then stained appropriately and the results studied under a scanning microscope.

The information obtained in this way can tell how long teeth will last before they are shed and replaced by new teeth. Plant-eating dinosaurs, who had to rip away at the hard grasses, had teeth that lasted only two or three months. Carnivorous dinosaurs, which, in general, ate softer food, had teeth that lasted as much as three years. Compare this with human teeth, which may remain in the mouth and functioning for decades. Of course, human beings can't replace teeth as reptiles can.

In any case, this knowledge that teeth offer can differentiate between old dinosaurs and young ones. It therefore can give one an idea on such things as birthrates. It will also help in estimating the number of predators versus the number of prey.

All this is not bad for information from an old, rocky tooth that may easily be 100 million years old.

Suzanne G. Strait of Duke University in Durham, North Carolina, has tackled teeth from a different angle. She has worked mainly with the teeth of small mammals such as bats and primates. She studied the enamel scratches on teeth and, working with living animals, showed that a diet of hard objects such as beetles and bone produced scratches different from those produced by a diet of soft objects.

The information thus gained was used to study the fossilized teeth of small animals. It showed that a group of early primates lived primarily on hard insects. This sort of knowledge is useful in trying to work out the evolution of the human species.

These are examples of the way in which completely unlikely sources can be cajoled into yielding information that can perhaps be obtained in no other way.

Bony Heritage

Now that its history has been pushed back 40 million years, a celebration of bone is due. You might begin the festivities by exercising, not just to build showy muscles, but to strengthen your bones. In conditions of bone loss, like osteoporosis, human life can be miserable. Keep your bones hard, and remember that the slogan in nature has not been "loosen up" but "harden up."

Life began soft and lived only in the sea. We know about it because some soft bodies, even one-celled creatures, remain visible to us as impressions left in fossilized mud. The "softness" of early life is not total, for without a firmer boundary to separate it from the surrounding sea, a cell would dissolve in its watery environment.

Cells solved the problem by constructing organic macromolecules that form membranes. Plant cells made cellulose from long chains of glucose molecules; animal cells made other macromolecules, mainly protein. With cellulose, plants could conquer land and grow as big as a redwood tree.

Animals used organic macromolecules in many ingenious

ways. Keratin is prominent in nails, claws, and hooves. Arthropods cover their bodies with chitin, a polysaccharide similar to cellulose. Thanks to the protection of chitin, arthropods invaded the land along with plants, but stayed relatively small because an exoskeleton limits size.

There was a revolution at the beginning of the Cambrian period, 570 million years ago, when animals began to add inorganic material to their bodies. Members of the phylum Mollusca did it by secreting shells from their body surface, or mantle. This hard surface tends to limit not only size but mobility. Giant squids can be that big because the shell has been reduced to a horny "quill" inside the mantle.

About 550 million years ago, the phylum Chordata began its triumphant history (we say "triumphant" because we think of ourselves as the best and brightest chordates of all). Down the backs of chordates are a hollow nerve chord and (at least in the embryo stage) a notochord made of collagen. This notochord was the forerunner of the remarkable internal skeleton that became the specialty of higher chordates like the subphylum Vertebrata.

Not all vertebrates have bony skeletons, but all have vertebrae enclosing the spinal nerve cord. Until recently it was thought that the earliest bone was in the form of protective bony plates, especially in the head region, at a time when vertebrae were made of collagenous cartilage. Sharks still possess a cartilagenous skeleton, but they have bony teeth and are thought to be descended from small early vertebrates with bony plates, now present as spines in some sharks.

True bone is a living tissue unlike the shells of molluscs. Forty-five percent of bone is mineral, 25 percent is water, and 30 percent is organic. The membrane around bone is called the "periosteum"; that inside bone the "endosteum." Cells called "osteocytes" form a lacy pattern through the mineral matrix, which is permeated by Haversian canals containing blood vessels and nerves. In most mammals, including humans, the central canals of long bones are filled with a blood-forming marrow.

Only with firm, bony skeletons could vertebrates conquer the land. No shark has emerged from the water. With strong bones, land vertebrates could get as big as dinosaurs or elephants. With hollow bones to reduce weight, birds can fly.

When and with whom did bone actually begin? Recent discoveries indicate that bone began as long as 515 million years ago in small, soft-bodied chordates called "conodonts" (named for their "cone teeth"). Using the scanning electron microscope and interference contrast microscopy, British scientists have shown that the conodont feeding apparatus is true cellular bone.

The important thing is that cellular bone is the unique possession of the subphylum Vertebrata. This means that conodonts are probably the earliest vertebrates—40 million years before the bone of other vertebrates appears in the fossil record. Conodonts also upset the theory that the earliest bone was for protection, because the possession of those teeth marks them as predators.

Human beings are pushing their environment to the limit and looking for living room. Can we—with our bony heritage—survive in space settlements? Stanford University professor Dennis Carter, a specialist in biomechanical engineering, has made mathematical formulas for different kinds of bones. In a computer program, the formulas help predict what bones will do in different environments.

Carter believes that bone formation may be guided more by environmental stresses than by our genes. A chief environmental stress is gravity.

Perhaps this means that if we humans want to live away from Earth, we'll have to terraform a planet at least as big as Mars, and make orbital settlements that rotate. Or invent a starship *Enterprise*, with artificial gravity!

Dinosaurs

Why were dinosaurs so large? They were the largest land animals that ever existed, some of them being ten times as heavy as an elephant. Two biologists, James Spotila of Drexel and Frank Paladino of Purdue, have tried to answer that question by studying turtles.

Reptiles like turtles are "cold-blooded." That doesn't mean they're always cold to the touch. If they stay out in the sun they warm up. If the temperature drops, however, they lack any biological mechanism to keep themselves warm, so they cool down. Birds and mammals have such a mechanism and are therefore "warm-blooded," remaining warm to the touch even in cold weather.

The warmer anything is, the more rapidly chemical changes can take place in it. When an animal is warm, it is lithe and active; when it is cold, it is sluggish and inert. Birds and mammals are active even in the coldest weather, but reptiles and other cold-blooded land animals move more and more slowly as the temperature drops, and, if it drops below freezing, they are likely to die.

Yet there is evidence that many dinosaurs led active lives and may have lived in cold climates, too. Is it possible that dinosaurs, or some of them at least, were warm-blooded? Some scientists think so.

On the other hand, it is possible for cold-blooded animals to stay warm even in cold weather if they are large enough. The source of animal heat lies in the chemical reactions that go on in living tissues. The larger and the heavier the animal, the more heat it produces in the course of ordinary living. The heat that is produced is lost to the outer world through the animal's surface area, and the larger the animal, the more surface it presents to the world.

The two properties, heaviness and surface, don't increase at the same rate as the size of an animal grows, however. The weight of an animal increases as the cube of its size, and the surface area only as the square. In other words, if you were to suddenly double all the dimensions of a particular animal, its surface area would increase by 2 × 2 or 4 times, but its weight would increase by 2 × 2 × 2 or 8 times. If you tripled its dimensions, its surface would increase 3 × 3 or 9 times, but its weight by 3 × 3 × 3 or 27 times.

For that reason, a large animal loses a smaller fraction of its body heat in a particular time than a small animal does. If a cold-blooded animal is large enough, the body heat it generates and the heat from sunlight it gains during the day can keep it going through the cold of the night and let it remain active at a time when smaller cold-blooded animals must hibernate and lie inert.

Is that the secret of the dinosaurs' size? Did they evolve into giants as their way of staying warm and active?

We can't measure dinosaurs' temperatures, unfortunately, for they are all gone, but what about large cold-blooded animals today? The largest cold-blooded reptiles still alive are estuarine crocodiles and leatherback turtles. Both can have weights of up to a ton, while the giant squid, which is the largest invertebrate, may be up to two tons in weight. (This is but a small fraction of the weight of large dinosaurs, but it is the best we can do.) Of these three creatures, the leatherback turtle is the most easily studied.

Leatherback turtles, swimming in the cold sea, seem to have body temperatures up to thirty degrees warmer than the water. Are they partly warm-blooded? If so, the rate at which they consume oxygen must be higher than that of small reptiles because it takes a lot of oxygen to bring about the chemical reactions that keep an animal warm.

Spotila and Paladino studied these large turtles in Costa Rica during the egg-laying season, when they came out on land. They measured the oxygen and carbon dioxide in the turtle's breath and found that it did consume oxygen more rapidly than other large reptiles. On the other hand, the rate of the turtle's oxygen consumption was less than half that of a warm-blooded animal of the same size.

The conclusion is that the leatherback turtle may have a way of generating more heat than would be expected but not enough to be considered warm-blooded. It maintains its temperature just by its size. The dinosaurs could have done the same thing.

Of course, being large has its disadvantages, too. Large animals reproduce more slowly and require more food individually, which means they must remain far fewer in numbers than is the case with small animals. That, in turn, means that if there is a sudden radical change in the environment, or a sudden decrease in the food supply, large animals are more likely to starve and die and even become extinct than small animals are.

Thus, when there was a cometary collision with Earth 65 million years ago and all sorts of disasters ensued, it was the large animals that suffered the most. All the dinosaurs and other giants of the time were wiped out, while the primitive birds and mammals, which were small and yet active because of warm-bloodedness, survived.

The Monster's Arms

New evidence has been obtained that takes up the matter of whether history's most ferocious land predator was really a terrifying monster or just looked like one. This comes about because two scientists, Matt B. Smith of Montana State University Museum and Kenneth Carpenter of the Denver Museum of Natural History, have been examining the forelegs of an allosaur.

The allosaurs were the largest and most fearsome-looking of all the meat-eating dinosaurs. The example best known is the one we call *Tyrannosaurus rex* (Latin for "king of the master lizards" and well-named if we go by appearance). The tyrannosaur was up to forty-seven feet long from the tip of its nose to the tip of its tail. It stood on its hind legs and its head reached a height of eighteen and a half feet, so that it was as tall as the tallest giraffe, but it was much more massive. Its weight was at least seven tons, which made it as heavy as a large African elephant. The most fearsome thing about it was its large head, about four feet long, with a large mouth, equipped with teeth that were over seven inches long and as sharp as butcher knives.

Most of us have seen a tyrannosaur in imagined action in Walt Disney's cartoon classic *Fantasia*, in which a fight between a tyrannosaur and a stegosaur is the climax of the representation of Igor Stravinsky's *Rites of Spring*. One did not have to be a child to be frightened by the tyrannosaur when it first appeared onscreen to a crescendo of music.

The tyrannosaur was built like a gigantic kangaroo, with a long tail and powerful hind legs. Of course, it was too massive to hop, but it could probably run, with a stride of some thirteen feet. As in the case of the kangaroo, the tyrannosaur's forearms were short, only three feet long, which is very small compared to its overall size. The arms are usually imagined to be more or less useless, merely writhing with fury as the tyrannosaur fought with its mighty fangs and the claws of its gigantic hind legs.

And yet some paleontologists feel that the absence of useful forelimbs limited the capacity of the tyrannosaur to deal with other large dinosaurs (like the stegosaur) that might be fighting for their lives. They suggest that the tyrannosaur, for all its appearance, was only a carrion eater; that it feasted on the leftover kills of other more agile predators, or else on kills by lesser predators who were small enough to be pushed or frightened away.

You might wonder why a carrion eater would have such ferocious jaws and teeth, but we have the modern example of the hyena. Hyenas are carrion eaters, but their jaws are extremely powerful. They can crush the strongest, hardest bones, but hyenas don't like to use those jaws on living victims. They skulk about in wait for other predators to do the initial work. Were tyrannosaurs outsize hyenas, or outsize wolves?

Smith and Carpenter studied a skeleton of a close relative of the tyrannosaur that was first uncovered in 1988. They examined the configuration of the bones of the forearm carefully and measured the width of the spot on one of the bones that marked where a tendon was attached, and to the tendon the biceps muscle.

From the thickness of the tendon, they decided that the biceps muscle must have been as wide as the human thigh and that the arm could have lifted a weight of as much as 426 pounds. Surely, arms like that are not useless.

Furthermore, each forearm had two claws that had the ability to flex extensively, which would not be so if the arms were useless. In fact, the two claws face away from each other.

Instead of facing each other, as in human beings, so that they could come together and grip, they seem to splay out.

Smith and Carpenter believe that the usefulness of claws of that kind is that each claw can impale a separate portion of the body of the victim. That would suffice to hold it helplessly while the jaws of the monster moved in for the tearing that would complete the kill.

If this is so, then the tyrannosaur was every bit as ferocious as it looked, and it would be doubtful if any other animal could stand up against the determined onslaught of a tyrannosaur. Once those deadly forearm claws plunged in, the fight was as good as over.

This discovery does not convince some paleontologists, who feel that the forearms, however powerful they might be, are too short to take much part in an actual fight to the death. They would still merely writhe furiously. In that case, why are they so powerful and so well-clawed?

There are those who say that once the tyrannosaur came across a kill, he used his forelimbs to get a good grip on the carcass while tearing it with his jaws—so that he was *still* a carrion eater. The matter, therefore, is not yet settled.

Dinowalk

Dinosaur is Greek for "terrible lizard," but everyone loves them anyway. Dinosaurs decorate T-shirts and are fast-selling toys, as well as the most popular museum exhibits. There's even a family of dinosaurs with their own TV sitcom, yet a live dinosaur has not been seen for 65 million years. Real dragons have never been seen, but they are equally popular. Perhaps there's a connection between the notion of dragons and, buried deep in our brains, a vague early mammalian memory of dinosaurs. Whatever, research on those extinct but very real dinosaurs goes on.

The Mesozoic was a spectacular era in Earth's history, starting 190 million years ago with the Triassic, when dinosaurs appeared as the descendants of primitive, "socket-toothed" reptiles called "thecodonts"—which also gave rise to pterosaurs and crocodiles. During the Jurassic and then the Cretaceous, dinosaurs were the undisputed masters of the world. Mesozoic mammals were small insectivores trying to stay out from under dangerous feet.

Dinosaurs came in all sizes, and in two main groups—the ornithischian (bird-hipped), and the saurischians (lizard-hipped). Many dinosaurs were much more stupid than any crocodile alive today, while others were brainier than any living reptile and, given the chance to evolve instead of becoming extinct, might have become more intelligent than we are.

Both groups of dinosaurs had quadripedal (four-footed) and bipedal (two-footed) members. The feet of large herbivores like the sauropods were elephantine to support the immense weight—from ten to thirty tons on the average, up to perhaps seventy-five tons in the biggest. Their legs were huge and placed under the body so they could walk like elephants and not have to wallow in marsh water to support their weight, as was previously thought.

New York's American Museum of Natural History has a new exhibit that shows a sauropod named *Barosaurus* rearing up on hind legs to protect her young from a predatory *Allosaurus*. With her long neck, she is so tall that only the height of the museum's rotunda ceiling accommodates her.

Bipedal dinosaurs ranged from chicken-sized creatures to enormous predators like *Tyrannosaurus rex*. One, named

Dromiceiomimus, looked a little like an ostrich and could run faster than a horse. Most bipedal dinosaurs had three-toed feet much like those of big birds, although some had more toes, and a few had fused toes similar to those of early horses (dinosaurs never got as far as evolving hooves). It used to be thought that hadrosaurs (duck-billed dinosaurs) were bipedal because their front legs were shorter than their hind legs, but now scientists believe that they could run on all four legs. Theories about specific groups of dinosaurs are under constant revision.

The prevailing opinion is that dinosaurs were not torpidly slow-moving and that many could run well. If they wished, they could run to the limits of their world until the unitary continent of Pangaea began to break up in the later Mesozoic. Further-more, good running requires good circulation. Robert T. Bakker points out that new data on certain dinosaur skulls affirm the theory that dinosaurs were warm-blooded, like birds. The skulls of birdlike dinosaurs and even the largest land predator ever, *T. rex*, have been examined with C-T scans, which reveal a nerve pathway like a bird's. The skulls also have birdlike air passages to keep the head light but also cool—needed if blood is warm.

Vertebrate paleontologist Emily Griffin examined the spinal canals in dinosaur vertebrae and concluded that most dinosaurs were indeed capable of an extensive range of move-ments, faster than expected. Studies of muscle marks on di-nosaur vertebrae show that dinosaurs had exceptionally strong muscles for land travel, like those of big, fast animals of today.

Bipedal dinosaurs were the fastest. Most of them had long tails for balance during running. Scientists know this because there are bony rods extending laterally down those tailbones, to hold the tail stiffly out while the animal runs fast. Some long-necked dinosaurs (especially small carnivores) also used their extended necks for balance in running. An interesting sidelight about some dinosaur neck vertebrae is that the di-nosaurs thought to be ground grazers all have S-shaped neck vertebrae, just as ground-grazing animals have today. Further-more, according to their bones, dinosaurs were not slow-growing like turtles and alligators, but grew as fast as birds do now.

Indeed, birds are considered by many to be feathered di-nosaurs, still alive in our time. The other day I sat on a bench in Central Park and noticed that, while sparrows and house finches hop, starlings and pigeons walk. When birds hop, their heads don't move back and forth. Look at a pigeon carefully. It can't walk without "head bobbing," which makes my neck hurt just to watch it. Apparently, there's no way yet of being certain that any dinosaurs hopped or that walking, running dinosaurs also bobbed their heads the way walking birds do.

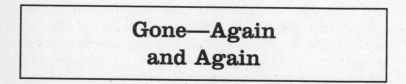

Gone—Again
and Again

Extinction is a nasty word, unless you want to inflict the pro-cess upon disease organisms, cockroaches, or the gremlins who bollix up your computer. Until the atom bomb was used in 1945, human extinction was generally considered only in religious terms, but a few years after the start of the atomic age, I heard a college professor give a lecture in nuclear phys-ics. His voice trembled as he talked about the Bomb, and our shaky future. I decided that if we became as extinct as the dinosaurs it would be our own fault. Since then I've come to realize that there are other possibilities.

We now know that there have been many mass extinc-tions, most of them quite mysterious. But not all. For one mass extinction, many scientists are convinced that they know what happened.

In 1980, Walter Alvarez suggested that 65 million years ago an extraterrestrial object (comet or meteor) hit Earth, causing such damage to the biosphere that many species, including the dinosaurs, became extinct. Sixty-five million years ago marks the "K-T" boundary, between the Cretaceous and Tertiary periods, and for years scientists tried to find a suitable impact crater that might account for the K-T extinction. Ten years ago a likely candidate seemed to be the 180-kilometer-diameter Chicxulub crater on (and just off) the north coast of the Yucatan peninsula, in Mexico. Finally, there's proof.

Scientists have found a chemical similarity between the once-molten rock in the Chicxulub crater and the glass beads (microtektites) discovered in Haiti and northeastern Mexico. Both the rock and the microtektites are the result of something hitting Earth with a powerful impact. The age of the crater has been accurately determined by means of a new geochronological technique called "argon-argon dating," which itself has been recently improved. The crater and the microtektites are the same age—65 million years.

Some geologists believe that there wasn't just one impact, 65 million years ago, but at least two—the known and dated Chicxulub crater, and another in Iowa. Perhaps whatever hit Earth broke up on its way down.

The K-T extinction was the worst in the past 200 million years, but there have been several others just as devastating to life on Earth, and far more puzzling.

Five hundred million years ago, in the Cambrian, many of the first hard-shelled creatures disappeared, cause unknown. Then, in the Devonian, 370 million years ago, the trilobites were decimated and 70 percent of other marine species became extinct. There may be some connection between the Devonian extinction and evidence of impacts near that time, but scientists aren't certain and some think that volcanic eruptions were the cause.

About 250 million years ago, the Permian extinction was far worse than the K-T extinction, for it killed off many land forms in addition to complex communities of life in the sea around Earth's one continent, Pangaea. Much of life was wiped out, for few fossils appear in the next years, during the early Triassic. Without evidence for a guilty asteroid impact, scientists postulate that the extinction was caused by a decline in oxygen levels, occurring when organic material is exposed and oxidized as sea levels drop and then rise again. Paleontologist Paul Wignall says that if we humans induce a greenhouse effect, raising sea levels, we might produce lower oxygen levels—and suffocation.

After the Permian extinction, life eventually started booming again, so that by the late Triassic, there were many species ripe for another extinction, which did occur. This one may also have been due to an asteroid impact, as evidenced by shattered quartz crystals found in Italy. Life picked itself up and, through the next periods of Earth's history, became even more spectacular. During the Jurassic and Cretaceous, dinosaurs of all kinds roamed the land, flew in the air, and swam the seas. Until the K-T boundary, and that big asteroid.

Although there hasn't been a mass extinction since the K-T boundary, Earth is continually being hit by small debris (even some of our own making). The most recent big impact was in 1908 when an object (a comet?) flattened trees in Tunguska, Siberia. We've all watched "shooting stars," which are meteorites burning up in our atmosphere. Some day something bigger and much more deadly could approach our planet, so people worry about how to prevent an extinction due to a large hit. The best idea is to do something about the object before it ever gets near Earth. A viable, global space program would help.

Not all large impacts have been deadly. Organic chemistry (and ultimately life) may have been helped, or even started, by "impact shock" when the young Earth was bombarded by material left over from the formation of the solar system. It's also possible that the bombarding matter, like carbonaceous meteors today, contained organic molecules, the building blocks of life.

Life, once it began, has been tenacious in spite of mass extinctions. As the Gaia people say, even if humanity becomes extinct, Earth as a living planet may survive whatever is done to it by extraterrestrial objects or by human stupidity.

Whales with Feet

Recently, paleontologists from the University of Michigan and Duke University, under the leadership of Philip D. Gingerich, uncovered the fossil remnants of an ancient whale in the Egyptian desert about ninety-five miles southwest of Cairo. What was a whale doing in a desert? Well, it wasn't a desert 40 million years ago; it was an arm of the ocean that has shrunk since and left behind the Mediterranean Sea.

What was really unusual about the fossil whale, however, was that it had two small hind legs that had the same kind of bones we have in our own legs, including the bones that indicated the presence of three toes on each foot.

The fossil record we find in the rocks is like an old and infinitely valuable book, which unfortunately has most of its pages missing. What's more, some pages that exist are too wrinkled and blurred to make out their contents clearly. After all, the fossil record is half a billion years old and has undergone enormous damage because of mountain building, land erosion, earthquakes, volcanic eruptions, and so on. On top of all that, most forms of life don't die under conditions that make for fossilization. The net result is that many puzzles remain in our attempts to make sense out of the fossil record.

For instance, mammals developed from reptiles on land as long as 200 million years ago. Since then, almost all types of mammals have continued to live on land. Some mammals have returned to a water existence in their search for food, but most such groups of mammals show clear signs that they have descended from land-living mammals.

Otters, for instance, are associated with rivers, but they are not very different from their land-living relatives, the weasels, ferrets, and the like. Sea otters are more thoroughly adapted to the sea, but their similarities persist, too. Seals, walruses, and sea cows are still more adapted, to the point where they have flippers rather than land-going limbs, so that while they are grace itself in the water, they are clumsy indeed on land. Nevertheless, their flippers have the same bone arrangements as are found in the legs of land mammals.

Whales and dolphins (the "cetaceans") are the real puzzle, for there the signs of land ancestry are dimmest. They are land mammals that returned to the sea 50 million years ago and are now the best adapted to water life of all the mammals. Cetaceans developed a streamlined fishlike shape, and while they have two foreflippers that were clearly once forelegs, they have almost no sign of hind legs. Deep within the flesh of their hip regions are small remnants that must once have been thigh bones, but that's all. Until now, no fossil whale has been discovered with more signs of hind legs than that.

Is it possible that whales are not mammals? No, that is not possible. They give birth to living young that develop inside their mothers by means of a placenta and that suckle on mother's milk after they are born. Whales have diaphragms, and in embryonic form they even show signs of having hair. They have all the characteristics of mammals, so it stands to reason that their ancestors must once have lived on land.

Unfortunately, the fossil record gave no sign of any mammal that had the characteristics that would lead us to say: This is a mammal that is on its way to evolving as a whale. Nor was any fossil mammal found that was clearly a whale but that showed more signs of land ancestry than modern whales do. It was an infuriating gap in the fossil record.

But this new fossil has been found to fill the gap in the record. It is a kind of early whale known as *Basilosaurus*— about fifty feet long, thinner and less bulky than modern whales. It has a relatively small skull, a small rib cage and a long and serpentine spinal column.

The *Basilosaurus* is about 40 million years old, so it was swimming the ocean 10 million years after whales had first evolved. And though 10 million years had passed, it still had hind legs. Not much, to be sure, but the bones are there and cannot be mistaken. If they were stretched out in a straight line, they would be two feet long. They include a femur (the bone of the upper leg), a tibia and fibula (the two bones of the lower leg), anklebones, and the bones of three toes.

That's very small in comparison to the size of the *Basilosaurus* itself, and the legs can't have had much use. They seem to have been permanently flexed and certainly could be of no use on land. They would be of little use in swimming, either. They might have been used to help scrabble out of mud in shallow water.

Some scientists postulate that the legs were primarily devices intended to hold the female during copulation. If so, it may be that they were useful only to male *Basilosauri* and are smaller, or absent, in females. It would be interesting, therefore, to find other examples of these fossils in order to examine that possibility.

Dead
as a Dodo

We now have a better idea of how the long-extinct dodo looked, thanks to a model prepared by A. M. Kitchener of the Royal Museum of Scotland in Edinburgh. He shows us a bird that is sleeker, more graceful than the one usually pictured. It now seems sadder than ever that the dodo no longer exists.

The history of man's relationship with the dodo begins in 1510 when the island of Mauritius in the Indian Ocean, due east of Madagascar, was sighted by Portuguese sailors. Although the island was known to Arab traders, it was still unoccupied at that time. It is a roughly elliptical island, 38 miles by 29 miles, and has an area of 770 square miles, about three-fourths the size of Rhode Island.

The Portuguese left it unoccupied, and the Dutch came across it in 1598 and named it for the executive head of the Netherlands, the *stadtholder*, Maurice of Nassau. The island received the Latin version of his name and became Mauritius.

The Dutch tried to settle the island and failed. It was taken over by the French in 1721. They fought the British over it, but it remained French, more or less, until 1965, when the island gained its independence. It now has a population of about a million, mostly of Indian and African descent. With all due respect to the people of Mauritius, however, the chief claim to fame of the island is a bird that no longer exists upon it.

The Portuguese were the first to report seeing the bird, after they had explored the island. It was a large bird, larger than a turkey, and weighed about fifty pounds on its stout yellow legs.

The bird had grayish feathers that were white in some regions, with a tuft of white feathers for a tail. Its tiny wings were useless for flight. Its most remarkable feature was its head, which sported a black bill, with a strongly hooked reddish tip that was like no other beak of any bird in the world. This bird was a member of the pigeon family and is frequently described as a large, flightless pigeon. It existed only on Mauritius, though related species, called "solitaires," existed on the nearby islands of Reunion and Rodriguez.

The bird of Mauritius, having no particular enemies, had never developed methods of self-defense. Earthbound and helpless, it seemed simpleminded to its Portuguese discoverers, who gave it the name "dodo." (It is from the Portuguese *doudo*, meaning "simpleton.")

Once Mauritius was finally settled, the settlers killed the dodo freely and so did the animals that came along with human beings. There was no feeling at the time that rare species ought to be preserved, no system of zoos that could save some animals when they disappeared from the wild. By 1698, the last dodo was dead, and a really magnificent and unusual organism was forever gone. The related species on neighboring islands were also wiped out within a few decades after that.

Now the dodo exists only in the common phrase "dead as a dodo." The word is also used to signify any person who is hopelessly behind the times and who finds refuge in blind conservatism. (Poor dodo! To be remembered for that.)

We know of the dodo's appearance from the diagrams that were made of it, from a few skeletons that exist, from one preserved head and a couple of feet.

We know its appearance best from the fact that a dodo appears as a character in chapter 3 of Lewis Carroll's *Alice in Wonderland.* John Tenniel, the famous illustrator, had it appear in two of his drawings, especially the one in which the dodo is handing Alice a thimble (her own thimble) as a prize. In the illustration, the dodo is pictured as an obese bird that we can easily picture as waddling about in ungainly fashion. Its behavior would thus match its name and help explain why it is extinct.

Kitchener does not think that is an accurate picture. It may have been based on some captive birds that were overfed and kept inactive. It may also have been influenced by the common statement that it was larger than a turkey, so that one thinks of domesticated, overfed turkeys. (Who knows! If dodos had been saved, we might now have dodo farms and dodo meat that might be superior to that of the turkey.)

Instead, Kitchener turned to earlier drawings that showed thinner birds. His model seems far less ungainly and could probably run in reasonable fashion. There is, after all, no use in adding the insult of ungainliness to the injury of extinction.

As an example of the interrelationship of species, there is a tree on Mauritius whose seeds will not germinate unless the fruit has first passed through the digestive track of the dodo. The dodo's digestive juices scarified the seed, and when it was finally deposited (with fertilizer), it would sprout. All the trees of this type on Mauritius are now at least three hundred years old. No new ones will sprout, and, eventually, the tree will be as dead as—well—a dodo.

The
First Catalyst

Another view as to how life may have originated is based on the work done, independently, by Sidney Altman of Yale University and Thomas R. Cech of the University of Colorado, who shared the 1989 Nobel Prize in Chemistry.

Here is the main problem that has been puzzling biochemists about the origin of life. There are two important sets of compounds in all living cells. There is a kind of nucleic acid called DNA, which efficiently stores all the genetic information and makes vast numbers of molecules just like itself to carry the information from cell to cell and from parents to offspring. DNA is in the cell nucleus, but it passes on the genetic information to another kind of nucleic acid called RNA. The RNA can go outside the cell nucleus, and there it can supervise the production of numerous protein molecules.

The protein molecules, in large numbers, each have a unique surface, and on that surface certain chemical reactions can take place speedily that ordinarily would take place only very slowly. The proteins act as "enzymes" or "catalysts," speeding and controlling the chemical reactions that make it possible for cells and organisms to carry through all the complex changes that make them alive.

Now here's the catch. DNA molecules can store genetic information with remarkable efficiency, but they cannot act as catalysts. Protein molecules are remarkable catalysts, but they cannot store genetic information—and a living cell has to be able to do both.

How, then, did life get its start? Did some DNA molecules form by the random interplay of atoms and molecules? In that case, those molecules could store genetic information and make additional molecules just like themselves, but those molecules, by themselves, couldn't *do* anything. Did some protein molecules form by the random interplay of atoms and molecules? In that case, they could catalyze reactions, but they couldn't control the production of more molecules like themselves and they would die out.

Well, then, did DNA molecules and protein molecules form simultaneously by random interplay? That's asking too much of coincidence, perhaps. Scientists have therefore been trying

41

to figure out how you can start with DNA molecules and develop proteins from them, or how you can start with protein molecules and develop DNA molecules from them, and no one has been able to work out a plausible scenario of either kind.

As it happens, though, when DNA passes on the information to RNA, it includes a lot of nonsense sequences (why they exist we don't know), and these get cut out of the RNA. Thomas Cech assumed that the nonsense sequences are cut out by the catalytic effect of certain proteins, since it was felt that only proteins were catalysts.

In 1982, he tried to isolate the particular protein catalyst or enzyme that did the sequence cutting. Little by little, all the enzymes were removed from the mixture in which the RNA was being purified of its nonsense—and the purification kept right on going. Finally, there was nothing in the solution, but the nonsense sequences were cut out anyway.

The only conclusion was that RNA itself had catalytic abilities. It could purify itself.

However, this was only the first step. So far, RNA was seen to work on itself but on nothing else. Then, in 1983, Altman found that another kind of RNA, called "transfer-RNA," also had to be purified of nonsense. This, too, was done by ordinary RNA and not by proteins.

The work was so carefully done that it was accepted by scientists almost at once. Continuing investigation showed that RNA molecules could catalyze a wide variety of chemical changes, and such molecules came to be thought of as a kind of enzyme. Since RNA stands for "ribonucleic acid," the catalytic RNA molecules came to be called "ribozymes," and Cech and Altman shared the Nobel Prize.

Now, you see, there is a possibility of visualizing the start of life in a way that avoids the DNA-versus-protein stalemate.

Suppose that sometime during Earth's first billion years of existence, RNA molecules were formed by the random interplay of atoms and molecules, bathed in the energy of sunlight, or lightning, or volcanic action. The RNA molecules could store genetic information and make more molecules like themselves. The RNA molecules could also catalyze various reactions, so that a kind of primitive "RNA-life" would form.

However, RNA molecules are not perfect. Small chemical changes would convert some RNA molecules into DNA molecules. These would not have catalytic effects, but they would store genetic information much more efficiently than RNA. RNA molecules could also form proteins, which would not store genetic information but which are far more efficient catalysts than RNA is.

As a result, a more advanced life made up of DNA and proteins would be formed and would oust the primitive RNA life. —Not entirely, though. RNA molecules still exist in modern cells and still have functions to perform.

The
Fifth Reptile

In 1989, Susan F. Schafer of the San Diego Zoo reported that a certain animal seemed to exist in three different species. If this turns out to be correct, it should be of devouring interest to zoologists, because the animal is one of the most unusual on Earth.

It is a reptile, and 100 million years ago the reptiles were the dominant form of land life. The huge and imposing dinosaurs, the large ichthyosaurs and plesiosaurs of the sea, and the pterosaurs that flew in the air were all reptiles. All of them disappeared about 65 million years ago, probably as the result of the collision with Earth of a sizable asteroid or comet.

Primitive birds and mammals survived, however, and so did some reptiles. The reptiles that survived the catastrophe and that still live today include four "orders" that are familiar to all of us. First, there are the turtles, which are the most ancient of the lot, which evolved even before the dinosaurs did, and which are still going strong. Second, there are the alligators and crocodiles, which are the closest living reptilian relatives of the dead-and-gone dinosaurs. Third, there are the various lizards, and fourth are the reptiles that have evolved most recently and are the most successful reptiles of the modern world, the snakes.

But wait, there is a fifth order of reptiles that hardly anyone but specialists has heard of. Over 200 million years ago, when reptiles were evolving into all sorts of varieties, there was an order called the *Rhynchocephalia* (RIN-koh-seh-FAY-lee-uh), which is from Greek words for "beak-heads." In fact, the order is sometimes referred to as just that, the beakheads.

The beakheads developed into a large variety of different species, some of them quite large, but it was not a successful order. Even as the dinosaurs arose and multiplied, the beakheads dwindled until only one genus remained alive, *Sphenodon* (SFEE-noh-don, from Greek words meaning "wedge-shaped teeth").

The animals of that genus managed to hang on, however, and when the catastrophe came that destroyed the dinosaurs, the *Sphenodons* survived somehow, and one species is still alive today and is found in New Zealand. It is called *tuatara* (TOO-hu-TAH-ruh, from Maori words for "spineback").

It looks like a large lizard, reaching a length of two feet, and sometimes lives for as long as a hundred years. It is gray, covered with white and yellow specks. But though it looks like a lizard, it is *not* a lizard. For one thing, it has a line of spines along the crest of its head and its back, which lizards don't have. It has a third, transparent membrane on its eyes, which lizards don't have. Its bones have certain features that are not found in lizards.

44

Perhaps the most fascinating thing about it is that there is an opening at the top of the skull under which lies the pineal gland (a portion of the brain). The pineal gland seems to have a structure resembling that of the eye (in the *tuatara*, it is sometimes called the "pineal eye"), and it may have a certain sensitivity to light. The resemblance to an eye is very marked in young *tuataras*, though in adult animals the skin of the head becomes pigmented so that little light can get through. The pineal eye may have helped the animal to determine the light level in the sky, distinguishing between sunny days and cloudy days, and between morning, noon, and evening, and guiding the animal's behavior accordingly.

In earlier times, the *tuatara* was found all over New Zealand. These islands had split off from other land masses so long ago that no native land mammals ever developed there, so that the *tuatara* (and various birds such as the giant moas) could exist in peace. Eventually, though, human beings and their domestic animals arrived, and the *tuatara* and other native New Zealand animals dwindled.

There are very few *tuataras* left, and the New Zealand government protects them zealously in order to keep the species alive. There are about five hundred *tuataras* on North Brother Island, which is only about ten acres in size, with twenty *tuataras* apiece on Stanley Island and Red Mercury Island. Ms. Schafer visited the various islands and detected sufficient differences in the creatures in the different places to make it appear that they exist in three closely related species.

Why bother to try to save these animals? For one thing, there's a certain sentimental attachment to animals that are "living fossils": antedating the dinosaurs yet still surviving. Can we bear to kill them off callously?

We must preserve the variety of life. All of life depends for its efficient functioning on the interplay between species. Every species that disappears tears a hole in the web of life and makes survival less likely for all the others. We *must* preserve.

Wrong about the Viceroy

Two Florida biologists, David B. Ritland and Lincoln P. Brower, have greatly weakened the popularity of a biological phenomenon known as Batesian mimicry.

The theory began with Henry Walter Bates, the son of a hosiery manufacturer who did not have much chance at an education before going to work in the hosiery business. Even though he had a thirteen-hour workday, he managed to go to school nights. Entomology, the study of insects, became and remained his hobby.

In 1844, Bates became friendly with Alfred Russel Wallace (who, along with Charles Darwin, later worked out the notion of evolution by natural selection). Bates got Wallace interested in entomology, and Wallace suggested a trip to tropical forests where they might collect specimens and learn something about the origin of species.

Following up this audacious scheme in 1848, the two friends landed in Brazil at the mouth of the Amazon. Wallace returned in 1852, but Bates remained for a total of eleven years, most of it in the virtually unknown upper reaches of the river. He collected over fourteen thousand animal species, mostly insects, more than eight thousand of which had not hitherto been known to Europeans.

Soon after he returned, Darwin's *Origin of Species* was published, and Bates accepted it wholeheartedly. In fact, Bates presented a great deal of information on insect mimicry, based on his Amazonian collection, that went a great way toward backing Darwinian ideas.

One cannot suppose that one insect species will deliberately imitate another in appearance. However, it is easy to see that imitations may arise through random variation. If the species being imitated is a harmful one in any way, or distasteful, so that predators avoid it, it turns out that the imitation is beneficial to the imitating insect. The imitator is avoided, too, and those that resemble the harmful insect the most are the least likely to be eaten. From generation to generation, then, it follows, that those that most successfully imitate the harmful insect survive the best.

This just fits Darwinian notions and is known as Batesian mimicry.

The best example we thought we had of Batesian mimicry is the case of the monarch butterfly and the viceroy butterfly. The monarch butterfly eats milkweed while in the larval form, and this gives to its tissues a baleful taste that no bird will ever attempt twice. A young bird who has never encountered a monarch may snap at it, thinking it a delightful morsel within reach, but one bite is enough. The bird flies away, obviously sick, and never touches a monarch butterfly again. (The monarch has a beautiful and very noticeable wing design so that it is easily recognized.)

I wrote an article about the monarch in 1990, and in it, I said:

> In fact, there is another butterfly, called the "viceroy butterfly," which is a bit smaller than the monarch, but is very similar in coloring, thanks to the blind forces of evolution that dictate that the ones that most closely resemble the monarch are the most likely to live long enough to reproduce. The viceroy is perfectly edible, but any bird that has tried to eat a monarch won't go near a viceroy either. No use taking chances.

I was wrong there. But I don't feel disgraced by that, for, apparently, all of science was wrong, too. Everyone was so convinced that the case of the viceroy was one of Batesian mimicry that no one ever bothered to test it. The two Florida biologists did, however, test it by taking three types of butterfly, the monarch, the queen, and the viceroy, and stripping off their wings so that the birds could not identify them by appearance and avoid them. The plump and naked torsos were then fed to unsuspecting red-winged blackbirds, which snapped at them eagerly and went through a strong rejection period.

It turned out that the monarch, the queen, and the viceroy butterflies were all undesirable. All tasted lousy. It was not, then, a case of Batesian mimicry, and biologists at once began to think that such mimicry did not take place as often as they had thought.

In that case, though, why did one butterfly mimic another so closely, if it were not trying to crawl under the protection of distastefulness? Current thinking is that where three distasteful butterflies all imitate each other closely, it benefits all three. The birds recognize the wing patterns and stay away from all three. That means that a juvenile bird who snaps at any one of the three stays away, thereafter, from all three.

After all, it might pick up a viceroy first off, and if that tasted fine, it would go for a monarch when it saw one. It would pay for all three to taste bad.

Ants—and the Animal Kingdom

The Bible considers ants to be examples of industry and forethought, working constantly and putting away food for the winter. It says: "Go to the ant, thou sluggard; consider her ways, and be wise" (Proverbs 6.6). The popular fable of the ant and the grasshopper also contrasts the industry of the ant and the hedonism of the grasshopper.

Nor does science neglect the ant. Recently, there was even a "First International Symposium on Interactions between Ants and Plants." Without going into the mind-boggling complexities that mark the relationships between ants and the plants they feed on and exploit, we can consider some amazing things about them.

All of life is divided into about thirty broad divisions called "phyla." I say *about* thirty because biologists don't entirely agree on the details of such classifications.

The phylum most familiar to us is, of course, the one we belong to: Chordata. The chordates include all the animals with internal skeletons that resemble each other in certain fundamental ways. They include all the mammals, birds, reptiles, amphibians, and fish. Human beings, sparrows, snakes, frogs, and mackerel are all chordates.

To most of us, all other phyla might well seem to be dismissable. They include such things as bugs and worms and plants and germs. They're not the kind of things we picture Noah as having saved on the ark. When we see illustrations of the animals marching into the ark, two by two, they are chordates, almost every one.

Of course, if we stop to think of it, even bugs and worms are important—but how important? Well, there is one phylum called "Arthropoda," which includes crabs and lobster, mites and spiders, centipedes and millipedes, and insects. Anyone studying the different phyla would have to admit that the arthropods are at least as important as the chordates. And, in one way, the arthropods are far more remarkable.

Every phylum is divided into a number of species, which cannot interbreed. Human beings are one species of chordate, for instance, and they can't interbreed with any other species of chordate. There are tens of thousands of different species of chordates altogether.

It may seem, then, that chordates are an example of a

phylum displaying a wide variety of different types of species, but the chordates pale in comparison with the arthropods. There are at least a million different species of arthropods, far more than the number of species of all other life forms combined. And, in fact, we have by no means studied and described all the species of life on Earth, and most biologists feel that the undiscovered species are almost all arthropods. There may be as many as 10 million species of arthropods in existence on Earth right now.

And of the different kinds of arthropods, by far the majority of the species occur among the insects. And among the insects the most common are the beetles. There are 700,000 known species of beetles, and who knows how many more remain to be discovered.

Why so many insects? They are small creatures who give rise to new generations every year, and in vast numbers. Vast numbers of individuals and generations mean that the evolutionary process is enormously fast compared with the slow multiplication of chordates.

New varieties of arthropods are constantly being formed, and any extraterrestrial studying Earth might decide that beetles were its most important inhabitants, at least in terms of quantity and variety.

Beetles, however, are among the larger insects. What about insects that are smaller? In particular, what about the tiny ants? The ants are a group that is far less diverse than beetles. There are only 15,000 species of ants known, and while many more ant species probably remain to be discovered, they will never compare to the beetles in variety. (However, the total number of mammalian species—the warm-blooded chordates with hair, including ourselves—is only 4,237, so you see how much richer in variety ants are than mammals.)

However, suppose we consider not the number of species but the number of *individuals*. Scientists have studied small areas of forests and counted every insect. Ants, it seems, make up some 70 percent of all individual insects, whereas only 10 percent of the individual insects are beetles. To put it another way, if you imagined a huge scale into one pan of which you ladled all the ants there were, and into the other all the other kinds of insects there are, the two would balance. The weight of the ants in existence equals that of all other insects combined.

We can, in fact, imagine something far more dramatic. Consider those huge scales again. In one pan, you have shoveled all the ants in their countless trillions. In the other pan, you place all other animals, excluding only the insects. In go all 5 billion human beings, all the elephants, hippos, cattle, horses, rats and mice, ostriches and eagles, snakes, tuna fish, worms, lobsters, and so on and so on. No use. The ants outweigh them all.

The
Duckbill Platypus

In 1800, the stuffed skin of an animal arrived in England from the newly discovered continent of Australia. The continent had already been the source of plants and animals never before seen—but this one was truly bizarre. It was nearly two feet long and had a dense coating of hair. It also had a flat rubbery bill, webbed feet, a broad flat tail, and a spur on each hind ankle that was clearly intended to secrete poison. What's more, under the tail was a single opening.

Zoologists exploded in anger. It was a stupid practical joke. Some unfunny character in Australia must have stitched together parts of widely different creatures in order to make fools of innocent scientists. There was, however, no sign of artificial joining. Slowly, after decades, zoologists agreed a new creature had been discovered. Its scientific name is *Ornithor-hynchus paradoxical* (birdbeak, paradoxical). To the general public, however, it became the "duckbill platypus" (*platypus* meaning "flat feet").

It seemed to be a mammal all right. The dense coat of hair was sufficient. Only mammals have hair. However, it did seem to lay eggs, and the egg-laying machinery was very like that of reptiles.

It wasn't until 1884, however, that the actual eggs laid by a creature with hair were found. (Such creatures included the spiny anteater, another native of Australia and New Guinea.) Such egg-laying mammals were called "monotremes" (one-hole).

It wasn't until the twentieth century, however, that the intimate life of the platypus came to be known. It is an aquatic animal, living in fresh water. The bill of the duckbill platypus is really nothing like a duck's. The nostrils are located differently, and it is a rubbery structure rather than a horny one as the duck's is.

The water in which the platypus lives is invariably muddy at the bottom, and it is in this mud that the platypus roots for its food supply. It can also detect weak electric currents that help it find its prey.

When the time comes for the female platypus to produce young, she builds a special burrow, which she lines with grass and carefully plugs. She then lays two eggs, each about three-quarters of an inch in diameter, surrounded by a translucent horny shell. These the mother platypus places between her tail and abdomen and curls up about them.

It takes two weeks for the young to hatch out. The new-born duckbills have teeth and very short bills and feed on milk. The mother has no nipples, but milk oozes out of pore openings in the abdomen. The young lick at these pores and are nourished in this way. As they grow, the bills became larger and the teeth fall out.

Yet despite everything zoologists learned about the duckbills, there still remained the question: Are they mammals with reptilian characteristics or reptiles with mammalian characteristics? Since we can't tell from the living creatures, what about the past? We have fossils of various animals, but these fossils consist mostly of bones and teeth. Could anything be determined from these?

Well, all living reptiles have their legs splayed out so that the upper part just above the knee is horizontal. All mammals, on the other hand, have legs that are vertical all the way down. Again, reptiles tend to have teeth that all look alike, while the teeth in mammals are differentiated, with sharp incisors in front, flat molars in back, and conical teeth in between.

As a matter of fact, there is a fossil called a "therapsid" that has vertical legs and differentiated teeth, but is considered without doubt to be a reptile, because there are other differences. In all living mammals, the lower jaw consists of a single bone. In reptiles it is made up of a number of bones. The therapsid's lower jaw is made up of seven bones, but one is very large. The other six are small and crowded into the rear angle of the jaw.

Mammals also have a palate so that inhaled air is led over it to the lungs. This means that breathing is not interrupted except for a second or two during swallowing. Reptiles don't have this palate because, being cold-blooded, they don't need a steady supply of oxygen. Some of the later therapsids do have a palate, which seems to indicate warm-bloodedness and perhaps even a hairy pelt. They were well on their way to mammalianhood, but they all became extinct. The only therapsids that are alive are those that have developed full mammalian characteristics and are mammals.

But that still leaves the duckbill platypus and the spiny anteater. Giles T. MacIntyre of Queens College studied the trigeminal nerve. In all mammals the nerve passes through one cranial bone. In reptiles, it passes between two bones. In a young platypus, whose skull bones have not fused, the trigeminal nerve passes between bones. MacIntyre believes that this makes the platypus a reptile. The argument, however, continues.

The
Real Unicorn

Gunter Nobis, former director of the Alexander Koenig Museum in Bonn, Germany, has examined the bones that have been found in the ruins of the ancient palace at Knossos, in Crete, ruins first explored in 1894. He has come to some interesting conclusions.

The palace at Knossos had an incredible maze of rooms, and many people think it represents the "labyrinth" that, in the Greek myths, was constructed for King Minos of Crete by the legendary inventor Daedalus. The story goes that Minos's queen, Pasiphae, fell in love with a sacred bull and that of this guilty affair a monster was born, the Minotaur, who had a man's body and a bull's head. The labyrinth was built to hide the Minotaur, which then fed on the king's enemies until Theseus of Athens slew him.

The truth behind this dramatic myth is that the ancient Cretans did indeed value bulls highly. This is not surprising, since the bull is an obvious symbol of fertility, and in ancient civilizations, fertility had somehow to be encouraged. That kept the herds of animals numerous, the grain harvests plentiful, and the human population itself growing. By worshipping bulls, then, with appropriate rites, all these things could be brought about, it was thought.

It is for this reason that the Israelites in the desert formed their "golden calf" (actually, a young bull) as an object of worship. Jeroboam of Israel set up two of them for his subjects to worship. And, presumably, the Cretans worshipped bulls as well. They even played games with them. At least, there are beautiful Cretan paintings that show young men seizing the horns of bulls and somersaulting over their backs.

It is not surprising, then, that the bones from the old labyrinth, which were studied by Nobis, are those of bulls. Over 60 percent of them were identified as bullish, but not all were the same. Some, indeed, represented the kind of cattle with which we are familiar. Others, however, were distinctly larger, and are thought to be those of the "aurochs," a wild ox from which ordinary cattle may have descended.

The aurochs (from an old German word meaning "primeval ox") was black and was considerably larger than ordinary cattle, some standing six feet tall at the shoulder. They had large horns, curving forward, and must have been formidable creatures indeed. Obviously, they would have to be bred smaller and milder if any real use was to be made of them, and this was what was done.

It is thought that, in the Bible, the aurochs is referred to as, in Hebrew, *re'em*. This is translated in modern versions of the Bible as "the wild ox," and the aurochs is referred to, in the Bible, as an example of an animal that is powerful and untamable. In the King James Bible, the word is mistranslated as "unicorn," and it is this that gives rise to the notion that a mythical one-horned animal must have existed because it is referred to in the Bible. Not at all! The real unicorn is the aurochs, and it had two large horns.

The aurochs survived through ancient and medieval times.

The last known herd of them existed in central Poland and was wiped out in 1627. That is sad, for they were magnificent animals.

From the bones studied by Nobis, it would seem that the Cretans had herds of both ordinary cattle and of aurochs, and that both were kept by them for food, for religious sacrifice, for games, and for breeding.

The most interesting of Nobis's findings is that some bones were of intermediate size. It is possible that cattle and aurochs were interbred and that hybrid animals of the two existed. Such animals were formed in the ordinary course of nature when cattle and aurochs were kept together. Perhaps the Cretans found the hybrids to be useful and therefore encouraged the breeding of more of them.

Thus, a mule, which is a hybrid of a horse and a donkey, has some properties in which it is superior to either of its parents (it is stronger, and more intelligent, than either, for instance). Though mules are not themselves fertile and can't have offspring, mule breeding has existed throughout history for the sake of their superior usefulness in some ways. It may be that the cattle/aurochs hybrid had its value, too, and that the Cretans found them particularly desirable and kept them for special purposes.

Because they might have kept these hybrids in seclusion, it is possible that an air of mystery arose about them outside Crete. People might know that there was interbreeding going on, yet not understand exactly what the interbreeding consisted of. Naturally, the most dramatic tale would be the one most reported, repeated, and believed, and what would that be but a man/bull hybrid? Perhaps it was in this way that the tale of the Minotaur was born. It may be our last remnant of the aurochs, the real unicorn.

A
Different Horse

A rare variety of horse—in fact, the rarest—is being restored to its own particular wilderness, where it has not been seen for a quarter of a century.

This horse first came to the attention of Western naturalists when an explorer from the Polish province of the Russian Empire shot one in western Mongolia in the 1870s. He brought back the skin and bones to the St. Petersburg Museum, and there naturalists found that it was not exactly an ordinary horse but was a species of its own.

Since the explorer's name was Nikolai Przewalski (pronounced "per-zhe-WAHL-skee"), the animal was named "Przewalski's horse." Whereas the ordinary horse that we see about us—pulling our wagons, running our races—has the scientific name of *Equus caballus*, the new species was named *Equus przewalskii*.

What is the difference between the two horses? Not much. Anyone looking at a Przewalski horse would immediately judge it to be a kind of pony with a dull, grayish-brown color, coarse hair, and a skimpy mane. Going into the minutiae, however, it has its differences, and perhaps the most notable is that each cell in *Equus p.* has two chromosomes more than are present in ordinary, humdrum *Equus c.*

Przewalski's horse made its home in the Mongolian region, and in ancient times it may have been widespread there, but in modern times, it had been reduced in numbers to a small herd in perpetual danger of extinction. And, indeed, that extinction came, after a fashion, for in the 1960s the last Przewalski horse was seen in the wild.

57

Yet the extinction was not entire and ultimate. A number of Przewalski horses had been captured and placed in zoos, and they seemed to have no difficulty breeding in captivity. The result is that now, with not one horse of the species left in the wild, about a thousand are flourishing in zoos. An effort is being made to transfer some of these odd horses back to Mongolia and restore them to the wild.

You might wonder why it is so necessary to place them in Mongolia. Is the Mongolian environment particularly suitable to the Przewalski horse, and will it flourish nowhere else? That would be strange since ordinary horses live well throughout the world—but that's precisely why Przewalski horses can't.

To understand this, consider how species separate. Ordinarily, a particular species gives birth to other members of the species and retains its identity. Of course, there are always mutations, small changes in characteristics that take place randomly, so that no two members of the species are entirely alike. Interbreeding mixes up these mutations and spreads them through the species.

If, however, two populations of a particular species are separated and stay separated for a long period of time, each population develops mutations of its own. If the period of separation is long enough, so many mutations of different sorts take place in each population, the two become different species.

Thus, camels all descend from a common ancestor, but the camels of the Middle East and the camels of Mongolia have drifted apart. They are still both unmistakeably camels, but the former, the "Arabian camel," has one hump, while the latter, the "Bactrian camel," has two humps, shorter legs, and longer hair. They are different species. The llama of South America, separated much longer, has changed so that it's not even recognizable as a camel, but it is a relative.

In the same way, there are two distinct species of elephant, the Indian and the African. And there are also mammals such as the tapirs that are descended from a common ancestor with the elephants, but have so changed that the relationship is not easily apparent.

As two species separate, they go through a gradual series of interrelationships. Eventually, they are so different that they cannot interbreed and, in fact, have no impulse to do so. However, before such a stage of difference is reached, two species might still interbreed, but give birth to infertile young that cannot continue the mixed species. Thus, horses and donkeys can interbreed, but give rise to mules and hinnies, which are infertile.

If species are still closer, they can interbreed and produce fertile mixed breeds. When this happens, and when one of the close pair of species is far smaller in number than the other, then the smaller number melts into the larger, and it is gone as an independent animal. The species of larger number can take the genetic admixture without substantial change.

This is the case of the two horses. If a herd of Przewalski horses were within reach of ordinary horses, there would be interbreeding and the Przewalski horse would vanish. For that reason, the Przewalski horses will be placed in a Mongolian area where ordinary horses are not found and where they will not be permitted to enter. In that way, this different horse may be preserved in the wild as a unique species.

Heady Stuff

Homo sapiens is a name redolent with hubris, for we humans do believe we're the smartest creatures on Earth. For centuries we thought no other creature had anything in its head worth bragging about, which explains why it's always news when other animals are shown able to think better than we imagined.

We reluctantly admit that creatures with brain anatomy closely resembling ours are more intelligent than the rest of the animal kingdom (we happily assume that plants are not smart at all and never will be). The chimpanzee and gorilla can learn sign language, become abstract expressionist artists, and generally demonstrate that they are our closest relatives.

We brag about our dogs, who seem smart enough to accept and follow us as pack leaders. Since a dog is a mammal, its brain is somewhat like ours but of course not as gloriously big in the important areas (the ones we use for heavy duty thinking).

Not all intelligence depends on having a big cerebral cortex. Perhaps the large striatal section in the parrot brain helps them do well in IQ tests. They can count and identify objects, colors, and shapes accurately, as well as make up words to describe objects, the way Koko the gorilla does.

When dinosaurs became birds that flew better than flying reptiles like pterosaurs, a great deal of weight was sacrificed in order to be true masters of the air. Since it wasn't possible for a flying creature to develop the brain size of a primate, the brain stayed small but functioned more efficiently than mammalian brains. Birds have a small cerebral cortex but, per size, a bigger diencephalon. Perhaps this and their superb circulation make their small brains more effective. Lowly pigeons not only avoid being tripped over by hurrying urban humans but can classify objects and pick out relevant ones from scenes beyond human visual powers.

Many scientists investigate consciousness in animals. Some study the remarkable "intelligence" of the seething mass of hive insects, but most concentrate on our own major division of the animal kingdom—phylum Chordata (animals with backbones) and specifically subphylum Vertebrata. Why not favor higher mammals and the best birds when researching brain-power?

The answer to that question is that it's easier to study simpler brains. But don't imagine that no other phylum has any intelligence worth mentioning. Now and then phylum Mollusca makes headlines. It's a fascinating division of the animal kingdom, already much in use by researchers who can study nerve transmission using the giant axon from the squid.

There are perhaps 100,000 molluscan species living either in water or in wet land habitats. Molluscs first appeared back in Cambrian times, when true vertebrates were millions of years in the future. Most molluscs are bilaterally symmetrical with one end more or less a head; their innards are covered by a fleshy mantle; they move along on a muscular ventral foot. A lot of them manufacture shells for themselves.

Humans love to eat molluscs. Members of the class Pelecypoda are frequently downed whole and alive. Unless your dinner lived in polluted water, don't worry about it. Clams and oysters are an example of "retrogressive evolution" of the phylum, because they don't have obvious feet or heads and aren't brainy at all. In Johnny Hart's cartoons, his talking clams have legs and feet, but real ones don't talk or walk.

Phylum Mollusca contains a remarkable class called "Cephalopoda," meaning "head plus foot"—because the tentacles emerge from the head. Of the ten thousand species of cephalopods that once roamed primordial seas, only about seven hundred still exist. Cephalopods have eight or more arms surrounding the head; inside the sharp jaws is a "radula" that rasps away at prey; powerful mantle muscle controls a jet propulsion system that permits rapid movement. Compared to sluggish clams, cephalopods have a closed, efficient circulatory system with thin-walled capillaries and quick exchange of gases.

The nervous system of a cephalopod looks unimpressive, the small brain made from a few fused ganglia, but thanks to convergent evolution, it works like ours in two ways. First, it is hooked up with precise "equilibrium receptor systems"— one for sensing gravity, and one for angular acceleration, enabling the squid or octopus to perform complex maneuvers. Second, the cephalopod eye possesses an image-perceiving retina similar to although not identical with ours.

Octopi are ingenious at finding ways out of captivity, even if it means going *through* air for a while. They can also be taught things, like picking out objects. Recently, researchers at the Naples Zoological Station have tested them for the capacity to learn by watching others. The unsociable octopus confounded expectations by quickly passing a test after observing what other octopi were successfully doing.

Don't underestimate brains unlike yours. When a brainier computer comes in, its circuitry won't look like a human brain, but it might work almost as well. Or better?

The Aye-Aye

Lemurs are the most primitive of the primates, and most of them live in Madagascar. They have foxlike faces and, unfortunately, are, for the most part, losing their fight for survival. This is largely because the tree cover in Madagascar is being destroyed so that their way of life is being wiped out.

The lemur that is rarest and oddest of all is called the "aye-aye." It is the largest nocturnal lemur in the world, with a measurement of eighty centimeters from nose to tail. It has huge ears like those of bats, and it has incisors that grow continuously, as is true of rodents.

When they come to a small hole in a tree, they can tell whether there is a grub inside or not. The way it gets its food is to tap the wood of trees in order to tell whether there are grubs underneath. They are the only mammals that use this system to find food.

Undoubtedly, its large ears allow it to hear grubs, and its ever-growing incisors can gnaw through the shells and bark of wood to get at the grubs. What's more, the aye-aye has a particularly unusual adaptation. It has a middle finger that is very thin and very long. This middle finger reaches into a hole and gets out a grub or a beetle that it proceeds to eat.

Some people point out that woodpeckers do much the same thing, using their strong beaks to prize out items of food that they then get out with their long tongues. However, there are no woodpeckers in Madagascar, and so there is the theory that the aye-aye plays the role of the woodpecker.

Carl Erickson of Duke University has studied the aye-aye in order to find out exactly how it locates its prey. He made use of four captive animals and placed mealworms in holes in wood. The aye-ayes had no difficulty whatever. They located the mealworms without any trouble. What's more, they weren't using the obvious holes in which the mealworms were hidden. Erickson added additional holes in the wood, and the aye-ayes paid no attention to them. They went for the holes that contained the mealworms.

He also found that the aye-ayes would break open cavities, even some that were two centimeters below the wood surface. Once they were opened, there would be mealworms or the equivalent in the cavities. Erickson noted that the aye-ayes made use of their long, middle fingers to tap the wood. They kept their faces close to the bark so that the large ears could be used to hear the prey.

How does the aye-aye manage to tap? It may hear movements; the tapping may cause the worms to give themselves away. It may use the sense of smell to help along in the tapping. In any case, it gets the food, and the tapping is simply unique among mammals.

It is unfortunate that the aye-aye should be living on the edge of extinction. We are used to seeing large animals in that sort of trouble. There is the African elephant, various forms of rhinoceros, the Siberian tiger, and so on. All of their numbers are rapidly decreasing, and it won't be long before the only specimens of those animals will be in zoos.

The aye-aye, however, is small and harmless, and it should not be in the same situation at all.

Orangutans live in trees in Borneo and Sumatra, but those trees are being pulled down, and the orangutans are being beaten backward. Eventually, they simply will not be able to find a living space, and they will be found only in zoos. The panda is dependent on bamboo, and when the bamboo is gone, the panda will be also. The koala lives on certain eucalyptus trees. When they are gone, so will the koala be. In short, animals will exist only as long as they have the way of life that they are used to.

Apparently, the aye-aye is also dependent on trees in Madagascar, and when they go, so will the aye-aye. We will still have them in zoos, but we don't know how well they might multiply in zoos. The panda, for instance, does not multiply at all readily in the zoo.

So it looks bad for the aye-aye and bad for those of us who find the aye-aye a peculiar and interesting animal. After all, the aye-aye, as I've said, is the only mammal that gets its food by tapping on wood; it's the only animal with a long middle finger that can be used to get out grubs.

Why should we let it go? It is much more important that we do our level best to keep it alive and nonextinct.

We may be able to do it. In recent years, we have worked diligently to keep animals alive that seemed to be on the edge of extinction. Why should we not do the same for the aye-aye? I rather think we will and that the little animal will be allowed to live in the forests of Madagascar, protected by ourselves and given a life of its own.

Our Closest
Relative

Data has piled up to show that chimpanzees are more closely related to us than they are to gorillas. We human beings have tended to believe ourselves to be separate from all other forms of life. After all, we are hairless, have no tails, walk on two legs, have the capacity for reasoning, and, according to some, have an immortal "soul." Therefore, here we are, and all other living things are on the other side of a grand divide.

Even as the tenets of evolution came to be more and more accepted, and people began to understand that human beings developed from "lower animals," it still seemed certain that there was a great gap between human beings and those lower animals.

Nevertheless, there is no question that the apes show considerable resemblance to human beings. Of the four kinds of apes, the gibbon and the orangutan seem relatively far removed from us, and it was taken for granted that while the gorillas and chimpanzees resembled each other closely, both were considerably different from human beings.

In the early 1960s, Morris Goodman of Wayne State University compared the blood proteins of the three kinds of animals to see how they cross-reacted. Blood from a particular animal would react strongly with a different but similar animal, and less strongly with one that was less related. He found to his surprise that chimpanzee blood cross-reacted with human blood more strongly than it did with gorilla blood. This was the first indication that chimpanzees and human beings, together, were on one side of the divide, while gorillas were on the other.

Many biologists were reluctant to accept cross-reaction data and remained with the old system, but in the meantime biologists were slowly learning how to analyze for actual genes. Genes control the chemistry of a cell, and in two different species of animals there are different genes. The more closely related two species are, evolutionarily speaking, the more genes they share.

In 1984 at Yale, Charles Sibley and Jon Ahlquist took gene sequences from one animal and reacted them with those of another. The more closely related the two animals were, the more similar were their gene sequences, and the more strongly would those gene sequences combine with each other.

Again, the result was that the chimpanzee gene sequences were closer to those of the human being than they were to the gorillas.

Scientists argued, but eventually it became possible to identify strings of genes and show what their structure was. It was no longer necessary to study the manner in which two gene sequences reacted with each other. All one had to do was to obtain a sample of genes from one species and show the nature of their nucleotides (the building blocks out of which they were constructed).

This was done for a section of human genes and then for the same section of chimpanzee genes and of gorilla genes. It turned out that there was a 1.6 percent difference between human beings and chimpanzees in that particular section. The difference between gorillas and chimpanzees, in the same section, was 2.1 percent. Other studies showed the same findings.

Recent studies have dealt with mitochondria, little structures in the cell that control the formation of energy. They contain genes. A team led by Maryellen Ruvolo, of Harvard

University, studied a seven-hundred-gene section from mito-chondria, a sequence that oversaw the production of an enzyme called "cytochrome oxidase." They studied the same section of genes from chimpanzees and gorillas and found a 9.6 percent difference between chimpanzees and human beings, but a 13.1 percent difference between chimpanzees and gorillas.

More and more, people are beginning to believe that chim-panzees and human beings are on one side of a great gulf, while the other apes (and, of course, other forms of life) are on the other side.

But if chimpanzees share so many of their genes with us, why are they so different? Well, the total gene information in the human body is equal to a thousand volumes of a large encyclopedia. If the chimpanzee only differs by 1.6 percent, that means there are still sixteen volumes of the encyclopedia that are different for us. That is enough to allow them to be a different species.

Two Hominids, Two Diets

Between 1.5 and 2 million years ago, at least two types of "hominids" roamed the plains of eastern and southern Africa. They were creatures that walked upright and resembled human beings more than they did apes. One of them died out, and the other survived to become the ancestor of modern human beings. Recently, an archeologist from the University of Cape Town, in South Africa, Julia Lee-Thorp, advanced an interesting reason as to why that might have been.

Of the two hominids, one was *Australopithecus robustus* and the other was *Homo habilis*, and they looked very much alike except for certain differences in the structure of the skulls. Of the two, *A. robustus* was somewhat larger and sturdier, but not much more. *H. habilis* may have had a slightly larger brain for its size, but not much more. The physical differences don't seem to be great enough to make it possible to use those as the reason why *A. robustus* died out and *H. habilis* became our ancestor. What did, then?

Diet! Ultimate survival may have depended on what they ate, but how can you tell what these primordial hominids did eat?

To begin with, everything that is, or was once, alive contains carbon atoms, and carbon atoms come in two stable varieties (or "isotopes"): carbon 12 and carbon 13. Carbon 12 contains six protons and six neutrons in its nucleus, twelve particles altogether. Carbon 13 contains six protons and seven neutrons, thirteen particles altogether.

The chemical behavior of a carbon atom does not depend on its nucleus, but on the electrons outside it, and *both* types of carbon atoms contain six electrons. This means that carbon 12 and carbon 13 act the same way chemically. What one does, the other does; where one goes, the other goes. This, in turn, means that every bit of carbon we deal with contains a mixture of both, and the same mixture, too. For every ninety carbon 12 atoms, we will find one carbon 13 atom.

But there is a catch to this simple picture. Although the two types of carbon do the same thing, carbon 13, having that extra particle, is just a bit heavier, and moves just a bit more slowly. This means that in any chemical process, it may end up that carbon 12 is a very tiny bit more common than it ordinarily is, or a tiny bit less common, depending on the process. What's more, chemists have learned to analyze the carbon atoms with such precision that they can measure the ratio of carbon 12 to carbon 13 well enough to pinpoint these slight changes.

Now every type of plant absorbs carbon dioxide from the air and puts it through a vast series of intricate chemical processes that end up incorporating some of those carbon atoms in their tissues. It is not surprising that different kind of plants do this in very slightly different ways and end up with a different ratio of carbon 12 to carbon 13. Just a tiny difference, of course, but from the ratio, chemists can identify different types of plants.

When animals eat plants (or other animals), the carbon atoms go through relatively simple processes in switching from plant tissue or animal tissue, or from one kind of animal tissue to another. For that reason, the ratio of carbon 12 to carbon 13 may stay very much the same as it was in the plants (or animals) that were eaten.

Bones contain a protein called "collagen," which contains carbon atoms, of course, and they can be used for the determination of such ratios. The catch here is that as bones age, collagen is lost. Bones in tropic areas that are more than ten thousand years old can no longer be used for the purpose. Lee-Thorp looked for something longer-lasting, and that means teeth. The enamel of teeth is the hardest tissue of the mammalian body. It contains only very tiny quantities of protein, but what it has it holds tightly and almost forever. Bones that are 1.5 million years old may be out of protein, but teeth that old can still supply the necessary carbon 12 to carbon 13 ratios.

Theories are not certainties. Processes in the animal body *may* indeed introduce slight changes of their own, or changes may be introduced very slowly after death. Nevertheless, the teeth of *A. robustus* indicate that this particular hominid lived on fruits, nuts, and grasses. There is nothing wrong with such a diet in itself, but *H. habilis* seems to have been omnivorous—that is, willing to eat anything.

Now, any restriction in the diet makes survival more questionable. If you depend too much on particular types of food, you are at the mercy of the supply. If, however, you can eat anything in sight, it is not likely that *everything* will go short at once. Omnivorous animals, such as rats, pigs, and human beings, have an enormous advantage. It would seem that *H. habilis* had it over *A. robustus*, and that is why the former may have survived when the latter didn't.

Ostrich Eggs and Humankind

We use several methods to estimate the dates of prehistoric events—measuring the breakdowns of different kinds of radioactive atoms; analyzing the annual layers of sediment ("varves") at the bottom of shallow water; examining tree rings—but would you believe studying ostrich eggs? In 1990, a team of scientists headed by Alison S. Brooks of George Washington University announced that the shells of ostrich eggs could be used to determine age.

This could be extremely useful. The most widely known method of age determination—the breakdown of radioactive carbon 14—will yield reliable results only up to 35,000 years ago. The next most common method, the breakdown of potassium 40, only gives reliable results for more than 200,000 years ago. The gap in between, 35,000 to 200,000 years ago, may be filled by studying ostrich eggshells.

Ostrich eggshells can only be found in parts of the world where ostriches are common, but in prehistoric times the range of the ostrich was greater than it is now. Ostrich eggshells are found in widely diverse areas of Africa and even in China.

The eggshells are found rather plentifully because they were useful in prehistoric times. One ostrich egg can supply the equivalent of two dozen hen's eggs, so it was a desirable food resource.

Once the top is broken off, however, and the contents eaten, what is the use of the eggshell itself? The shell of an ostrich egg is remarkably strong. It is about one-sixteenth of an inch thick and is so designed for strength that a 250-pound man can stand on one without breaking it. In the days before pottery, the ostrich eggshell was a perfect vessel—light and strong—to carry water. One eggshell with the narrow top broken off could easily hold a quart of water.

Naturally, then, ancient sites of prehistoric settlements would contain these shell fragments in the place of the pottery that one would find in later remains.

How does this help with prehistoric date determination? Well, there is always some protein in material that is living or was once living, even in hard objects like bones or seashells or eggshells. The protein molecules are made up of chains of smaller units called "amino acids."

Amino acids can come in two varieties, labeled "L" and "D," that are mirror images of each other (like your right and left hands). When chemists form amino acids in the laboratory, both varieties are formed in equal quantities. In living organisms, only *one* variety, the "L," is formed.

Each variety is pretty stable, but if untouched for a long period of time, such as thousands of years, there is a slow tendency for some of the "L" to turn into "D." From the amount of "L" and "D" you find in a particular ostrich eggshell, you can tell how much time has elapsed since the ostrich laid that egg.

This technique has been used since the 1950s on old bones, for instance, but there is a catch. In warm temperatures, the rate of change speeds up, and moisture causes it to speed up, too. You can't always tell how warm temperatures have been in the past, or how much humidity and rainfall there have been, so you can't be sure the rate of change from "L" to "D" has always been the same. It may have been faster at some times and slower at others. That introduces a considerable uncertainty into the actual age of an object.

Recently, however, this method of measuring age has been considerably refined. In addition, the ostrich eggshells are much less porous than bone. Water does not get into them as much, so that it seems reasonable to suppose that they are less affected by humidity and rain and their ages can be more reliably determined.

As a result, the team of scientists reporting this finding believe that they can determine the ages of prehistoric sites in the Kalahari desert with fair accuracy. The ostrich eggshells found in the older layers of sediment there may be between 65,000 and 85,000 years old. It seems possible that ostrich eggshell remnants might be found that are up to 200,000 years old at some places in Africa. In China, where the temperature is, on the whole, cooler than in Africa, it is possible that ages of up to 1 million years can be determined.

It is, of course, very useful to know how old relics might be, but, in particular, we might now be able to come to some better conclusion as to how old "modern man" (a term that describes all the people that live on Earth today) might be.

The general feeling is that modern man, *Homo sapiens*, first appeared 50,000 years ago, but ostrich eggshells are sometimes found in conjunction with human bones that seem to be those of modern man. If we determine the age of the shells, we will have the age of the bones. We can then estimate *when Homo sapiens* appeared, and even, perhaps, sharpen our notions as to *where* we began.

Crossing
to Australia

When did human beings first reach Australia? The oldest fossil human bones found in Australia date back about 30,000 years, and the usual estimate is that human beings, the ancestors of the present-day Aborigines, arrived about 40,000 years ago. However, in 1990, a group of Australian scientists, led by Richard Roberts, have advanced evidence that the first human beings may have reached Australia as much as 60,000 years ago.

At a site in northern Australia the group uncovered stone objects that look like human tools, although no actual relics of human bodies were found. By studying the manner in which grains of quartz at the site give off light when heated ("thermoluminescence") they can tell how long those grains (and the tools associated with them) have been buried.

This possible early appearance of human beings in Australia raises interesting problems. The ancestors of *Homo sapiens* ("modern man") were the smaller-brained *Homo erectus*. The remains of *Homo erectus* have been found in the large land mass of Eurasia and Africa, but not in either the American continents or in Australia. *Homo erectus* apparently could not cross the water barriers that lay between Asia and either North America or Australia. Perhaps the Siberian winters also defeated them.

A famous fossil remnant of *Homo erectus* was discovered on the island of Java in the 1890s. However, the island of Sumatra is separated from the Malay peninsula in southeastern Asia by only a narrow strait, and Java is separated from Sumatra by an even narrower one. *Homo erectus* may have drifted across on rafts, or, if there was a drop in sea level, as happens now and then, they may even have waded across. Between the western and eastern islands of what is now Indonesia, there is a wide and deep channel, however, that they were never able to cross.

Homo sapiens, more enterprising and with a larger brain, ranged out farther. Modern man was first to colonize Australia and the American continents. In the early stages of human history, the great Ice Age gripped the Earth, but Siberia was not covered by as thick or as extensive an ice cap as North America was. Hunters in search of large animals for food, the mammoth in particular, ranged farther and farther into Siberia.

So much water was tied up in the continental ice caps that the sea level was nearly four hundred feet lower about twenty thousand years ago than it is now. A land bridge connected Siberia and North America. Siberian hunters crossed it into North America and slowly spread out over the vast expanse of the American continents. When the ice caps melted and the ocean surged back into the Bering Strait, the humans in the Americas were isolated from the rest of the world till the European Age of Exploration began.

But what about Australia? At the height of the Ice Age, Malaya, Sumatra, and Java were undoubtedly all connected. So were New Guinea and Australia. For human beings from Asia to get into Sumatra and Java, and perhaps even into Borneo and Celebes, would have been relatively simple. To cross from Celebes to New Guinea and Australia would have been difficult, however, for there would be no land bridge they could use.

If, indeed, human beings arrived 60 thousand years ago, the continental ice cover was far smaller, and the sea level was higher. A stretch of water, at least 250 miles across, had to be crossed. Is there a chance that the first people who crossed into Australia were *Homo erectus* or, possibly, Neanderthals, an early subspecies of modern man?

This seems unlikely. We have no evidence that either

74

Homo erectus or Neanderthal man ever possessed the capacity to cross wide bodies of water. Furthermore, the objects found at the North Australian site included a grindstone and ground-up red and yellow minerals. This suggests that the people who inhabited the site decorated themselves and the walls of their shelter. This is a *Homo sapiens* characteristic, and, indeed, the Aborigines were doing that when the Europeans first reached Australia.

The reasonable supposition, then, is that the first to reach Australia were *Homo sapiens* and that this was done very early in the history of modern man. Indeed, if all this is correct, it casts a surprising light on the Aborigines. Europeans have long thought of the Aborigines as the most primitive of the varieties of modern man, but if their early arrival in Australia is correct, they may well have been the first human beings to make use of rafts or canoes to cross a wide stretch of open water. Not bad for a supposedly "primitive" people.

Vital Cooperation

Cooperation is an important word for the living world. All multicellular creatures exist thanks to the cooperation of their component cells. Sometimes cooperation on the cellular level can seem amazingly altruistic. While sperm are ordinarily thought of as independent and competing, it's been found that a normal rat sperm (trying to be first to impregnate the egg) is aided in its purpose by other, usually deformed sperm in the same batch. When the normal sperm is on its way to successful impregnation, the deformed sperm clump together to form a plug that prevents other sperm from entering the reproductive tract.

Certain bacteria apparently cooperate to hunt and attack prey. Other bacteria, very long ago, entered living cells not to destroy them but to cooperate for mutual benefit. These bacteria became the mitochondria, essential parts of cells. We could not do without our own mitochondria.

It seems that self-interest, once thought to be the driving force behind evolution, is not precisely the norm. Being altruistic and cooperative has enormous advantages, ensuring that the species will survive even if the individual does not. This is most dramatically seen in the social insects—termites, bees, wasps, and ants. Some members of these complex societies never reproduce, but work to help others of their kind.

Genetics builds altruism into the bee, for instance. A female but sterile bee worker can't help being born that way and, without thinking about it, automatically strives to help gene-related sisters. In other notably cooperative animals, the cooperation is built up from the devotion of parents to offspring, some of whom stick around to help feed those parents and their own new siblings.

The African naked mole-rat belongs to the family Bathyergidae, suborder Hystricognathi (which also includes guinea pigs and porcupines). They look like the mainly hairless babies of real rats, but they are fascinating, for they are the mammalian equivalent of social insects. They have one queen, who reproduces, and altruistic workers who tend her and keep the large underground colony going. Their complex society was first described only a few years ago by Jennifer U. M. Jarvis and is now under intensive study. We humans seem to be intrigued by selflessness and cooperation that really works.

Primitive humans tamed the wolf and turned it into a helpful dog who considers a human his pack leader. This was possible because both the human and the wolf knew how to hunt and live cooperatively. Like humans, wolf pups are not genetically programmed to be part of a social group, but learn. A pup raised by humans considers them to be his "pack" and obeys the pack leader—all to our advantage. Lately, ranchers in the West have been using this aspect of dogdom by raising puppies with sheep. The puppies grow up thinking of the sheep as their family, which they defend.

Our closest relatives, the chimpanzees and gorillas, also live cooperatively. Fighting chimpanzees will "kiss and make up" afterward, to keep the peace. Forest chimpanzees have recently been found to hunt in large cooperative groups, much the way it's thought our hominid ancestors behaved. The forest chimps are more inclined to share meat than other chimps, and it's probable that early hominid survival was improved by such altruism.

Jane Goodall's work with chimpanzees shows that individuals survive longer when they form close, long-term bonds with others. One chimpanzee mother lived into old age because her son cared for her. Unfortunately, when she died, he apparently died of grief. Close cousins to us, indeed.

Human beings are social primates par excellence. It's not possible to gaze out of a window onto the towers of Manhattan without admitting that nothing like that can be built without cooperation, regardless of whatever human behavior might be going on in the streets below.

Recent evidence gathered by Stanford archeologist John Rick indicates that the Stone Age Peruvians lived more ecologically than primitives have been thought to do. To preserve breeding populations, they apparently tried to avoid killing young animals. They kept their own population stable without resorting to infanticide—probably by telling young adults to go elsewhere, and by using sexual abstinence. They lived in cooperative harmony not often known nowadays.

As nation-states bicker with each other or break up into more bickering splinters, it's difficult to view humanity as a cooperative animal. But human beings *are* socially cooperative animals from way back, and are part of a biologically cooperative planet Earth. When we aren't cooperative, the results are grim and easy to see.

Left, Right

Everybody's got it—handedness: the tendency to use one of your hands most of the time. Right-handed people's centers for speech and fine motor skills are in the left cerebral cortex. Left-handed people (about one person in ten) have cortical speech centers on either side of the brain or in both.

Lefties have been given a pessimistic press. Left-handed children, coping with a right-handed world, are more likely to be dyslexic and to stutter. Earlier studies seemed to show that lefties were more prone to various physical illnesses, to die sooner, and to have more accidents and injuries.

The original gloomy data are now being questioned. It seems that left-handedness was more likely to shorten life in people born before 1890, but not since. The tendency to accidents is a cloudy issue, too, for it depends a lot both on the age and gender of the person and on what kind of accident it is.

Left-handedness is not so bad. A favorite genius of all time, Benjamin Franklin, was a lefty, and he lived to a ripe old age, during much of which he fathered the birth and early development of the United States of America. J. S. Bach was a left-handed musical genius who fathered twenty children, three of whom became composers. Leonardo da Vinci wrote easily in mirror script, as can many naturally left-handed people.

Why are so many humans right-handed? Our primate relatives don't seem to favor either hand in their ordinary activities, but our earliest ancestors were apparently right-handed in daily life. Hominids living about 2 million years ago made stone tools by hitting one stone against another to create a sharp edge. A right-handed user flakes a stone clockwise so that the flakes have some of the rock's surface only on the right side. Early hominids obviously held their hammer stones in their right hands.

New research hints that right-handedness may have been evolving long before hominids. While it's true that other primates don't favor either hand in ordinary life, under some test conditions they do. They tend to use their right hands for fine motor skills, especially when it's difficult for them to see what they're doing. Scientists postulate that their tree-dwelling ancestors held onto branches with the right hand and reached out with the left. Perhaps when some primates became largely ground dwellers they used the stronger right hand to crack open nuts. Even now they tend to groom more with the right hand than the left.

The most interesting aspect of the handedness problem is that both humans and great apes hold their babies on their left side. This has been said to be evidence of an early bias toward right-handedness, caused by the fact that if the baby is held on the left, the mother hears it more with the left ear and sees it primarily in the left visual field. Left hearing and vision are processed by the right half of the cortex, which takes care of emotional sensory input. The idea is that the mother automatically holds the baby so information about it goes to the side of the brain best adapted for interpreting and responding to the emotional input. Perhaps this enables the mother and baby to form a better emotional bond, so necessary for infant survival, or perhaps babies are held on the left because they can hear the mother's heartbeat better there. It's true that the heart is actually in the center, but that's under the breastbone. The apex of the heart is slightly to the left, where a doctor listens to one's heart with a stethoscope. A baby spends nine months listening to mother's heartbeat and after birth does seem to relax better when carried on the left.

Which came first, right-handedness, speech, or bipedalism? Perhaps body hair came either first or with bipedalism. All other primates have body hair for babies to cling to. Gorilla babies, as helpless as ours at first, can cling to mother's hair at the age of three weeks. If a naked hominid tries walking on two feet (perhaps to see over the taller grass in the predator-rich savannah), the baby has to be carried by the mother. If a baby cries less on the left, both it and the mother will be safer from predators.

If right-handedness goes way back in our ancestry, keep in mind that, according to the latest theory, the universe itself may be left-handed. From the skew of nuclear particles to the amino acids in living cells to the way galaxies rotate, there is definitely a left-handed bias in the universe.

Genes in Action

Keeping up with what's going on in genetics is a herculean task, since the field is expanding so rapidly. This is a brief introduction and a few examples of current work in genetics.

To begin with, genes are not simple little building blocks of heredity. When the Austrian monk Mendel was carrying out his great experiments in the 1860s, he had no idea that chromosome bodies in the cell nucleus were responsible for his interesting results in crossbreeding pea plants. He kept careful records and noted that his pea plants had definite, measurable characteristics he called "factors," which seemed to come in pairs. Mendel found that a "recessive" factor was not destroyed but remained hidden and could come out in the next generation.

When chromosomes were named—by von Waldeyer in 1888—nobody knew their composition, but it finally became clear that chromosomes had something to do with Mendel's "factors." In 1909 Danish biologist Wilhelm Ludvig Johannsen named the factors *genes*, a Greek word meaning "to give birth to."

Since the number of chromosomes peculiar to each species is far less than the inheritable characteristics, biologists decided that each chromosome must be a collection of genes. Thanks to the easily raised little fruit fly with its meager four chromosomes, it was found that genes are arranged and linked in complicated ways. Furthermore, as Mendel had already discovered, genes can cross over and mutate. Animal breeders had long been taking advantage of the tendency of "germplasm" to mutate. In 1791, from one odd lamb, a Massachusetts farmer bred short-legged sheep that couldn't jump over the stone walls around his fields.

The fruit fly made it possible to count genes in a chromosome (a minimum of 10,000 in the fruit fly) and find out that the individual gene has a molecular weight of 60 million. Human beings have larger chromosomes than fruit flies. Ours contain from 20,000 to 90,000 genes per chromosome pair. The molecular weight goes up and so do the complications.

Scientists are grappling with the complications. Modern molecular genetics is awesome stuff, trying to understand and *control* how genes work. Those "building blocks of heredity" have turned out to be DNA molecules (the famous double helix). DNA is deoxyribonucleic acid and is composed of four different nucleotides in various combinations now called the "genetic code." Scientists are trying to alter the combinations, which gets publicity as recombinant DNA, or "genetic engineering."

Any point to this work? Sure. There are little children who have to live in sterile bubbles because they have genetically missed out on a good immune system. Some day genetic engineering may give them a normal life. Perhaps eventually people with damaged DNA can have it repaired somewhat.

There are ongoing medical experiments that attempt to save lives through gene therapy—the insertion of healthy or specifically altered genes into diseased or deficient patients. Transfusing genetically altered lymphocytes into people with lethal immune disorders shows some promise.

Diagnosis improves, too. A leading medical journal reported that it is now possible to examine genes in circulating lymphocytes of people from families prone to familial hypertrophic cardiomyopathy, a disease that makes itself known only after an affected child grows up. Perhaps this screening test will identify the disease before it starts, eventually make possible some sort of preventive measures, and—not least—reassure other members of the family who do not carry the affected gene.

As gene therapy progresses, soon there may be advances in the treatment of cancer, diabetes, and other problems that threaten so many lives with debilitating illness and death. These treatments do not turn people into "mutants" or risk the rest of us with runaway horrid diseases.

Work on genes isn't confined to the medical field. Some of it is downright charming. For instance, DNA from preserved tissue of the extinct quagga has been found to be identical to that of zebras alive today. The quagga was zebralike in front, plain brown in back with a white brush of a tail. Zebra breeders are now trying to produce animals that look like quaggas.

Tomorrow the quagga—next century the Neanderthal?

Genes, Oncogenes, and Cancer

A fundamental discovery in connection with cancer won the 1989 Nobel Prize in Physiology or Medicine for J. Michael Bishop and Harold Varmus of the University of California in San Francisco. The story of how it came about starts three-quarters of a century ago.

In 1911, an American physician, Francis Peyton Rous, reported that he could transfer cancer from one chicken to another. If a chicken had a tumor called a "sarcoma," he could mash up that sarcoma and pass it through a fine filter that would produce a clear liquid with no living cells in it. If he injected some of the liquid into a healthy chicken, that chicken would develop a sarcoma.

It seemed the liquid must contain a virus, an infectious agent so small that it could pass through the filter. This particular virus was called the "Rous chicken sarcoma virus," and the discovery implied that at least some cancers were virus diseases.

Other scientists were skeptical, but as the years passed additional cases of the production of cancer in animals by injection were found, and in 1966, fifty-five years after the discovery, Rous (then eighty-seven years old and still working) received a Nobel Prize for it.

It was known that viruses contain nucleic acid. So do all living cells. The important variety of nucleic acid in cells—whether human or chicken—is called DNA. DNA forms another kind of nucleic acid called RNA (ribonucleic acid), which supervises the production of all the various protein molecules in cells. It is the protein molecules, in great variety, that keep cells, and organisms made up of cells, working so marvelously.

That is the central dogma of biochemistry: Genetic information passes from DNA to RNA to proteins.

The tumor viruses, however, contain RNA. When they produce cancer, they must do so by modifying the DNA of the cells in such a way as to start a chain of changes that ends with cancer. Because, in this case, genetic information passes from RNA to DNA, the tumor viruses are called "retroviruses," the prefix "retro-" being Latin for "backward."

Until the mid-1970s, the general feeling among scientists was that somehow cells (human and otherwise) picked up these retroviruses from outside. The retroviruses remained quietly in the cells, doing nothing for the most part, resembling ticking time bombs. Then, sooner or later, some effect, whether radiation, chemicals, or something else, would activate them, and they would set the cell on the path to cancer.

A few scientists, however, thought differently. They felt that cancer did not depend on outside viruses, but was built into the normal workings of the cell itself. There were certain normal genes in the cell that were themselves ticking time bombs. It was these normal genes that could be affected by radiation, chemicals, or something else, and that would, as a result, change slightly and become abnormal genes that initiated the changes that led to cancer. The abnormal cancer-producing gene was an "oncogene" ("onco-" coming from a Greek word for "tumor"). The normal gene, preceding the oncogene, was a "proto-oncogene" ("proto-" coming from a Greek word for "first").

In 1976, Bishop and Varmus announced the first important experiments that seemed to demonstrate that this oncogene notion was correct. The proto-oncogenes helped control normal processes of cell growth and differentiation, multiplying the number and different kinds of cells. These normal processes come to a halt when the body has enough cells. When the normal genes are changed into oncogenes, however, they lose the ability to call a halt. Abnormal cells grow without limit, invading normal tissues, disorganizing the body, and, in general, producing a fatal cancer.

Since 1976, further evidence has backed the view. It turns out that retroviruses are also the products of the oncogenes and may also be formed within the body and need not come from outside. As a result, Bishop and Varmus received the Nobel Prize thirteen years after their crucial experiments.

Presumably having sharpened our knowledge as to the origin of cancer, scientists might labor to identify the proto-oncogenes, to study carefully the nature of the changes that form oncogenes out of them. The hope would be that appropriate treatment might prevent or, at least diminish, the chances of oncogene formation, or ameliorate (or even reverse) their effects once they have been formed.

But why should normal genes exist that can, at any time, undergo cancer? For one thing, life has existed for about 3.5 billion years, but it is only about 0.8 billion years ago that organisms made up of many cells formed. Perhaps evolution has not yet perfected the system of producing and managing a combination of many cells, and cancer represents a lingering imperfection of the process.

The
Beautiful Microbes

Human beings have not ordinarily considered microbes to be beautiful, or even tolerable. When microorganisms were first seen by Anton van Leeuwenhoek in the seventeenth century, no one was particularly interested. After Louis Pasteur's germ theory of disease in the 1860s, people knew that microbes existed and that they were all too often nasty to human life.

The microscopic organism most condemned is the lowly bacterium. Its name comes from the Greek word for "staff," perhaps because so many bacteria resemble little rods. The biblical psalmist was certainly not talking about bacteria when he said that famous line about thy rod and thy staff they comfort me.

Many bacteria, however, are comforting to human life. The best-known "good" bacteria are those that help make soil fertile by fixing atmospheric nitrogen in it and those that live in the human intestinal tract to help digest food. And if it weren't for bacteria, we'd be skyscraper-deep in garbage.

Bacteria are newsworthy of late. The molecular biology of behavior is being studied with the aid of *Escherichia coli*, normally found in human intestines. *E. coli* moves along by rotating its helical filaments clockwise or counterclockwise at alternating intervals. Apparently *E. coli* can measure the change in the amount of chemicals in its environment and change its own motion to suit, which fascinates scientists.

A new Japanese deep-diving submarine is going to bring back microbes from the deep ocean bottom. While studying deep-sea ecology, scientists will be trying to learn more about the early evolution of life. It is thought that marine hyperthermophiles (microbes found in undersea thermal vents) may be the ancestors of all living creatures. Biotechnologists (a thriving group in Japan) will also search the deep-sea bacteria for genes as useful to them as those found in bacteria that live in hot springs.

Many other microbes are already in use by the biotechnology industry. One has helped make a Japanese detergent, because its "cellulase" enzyme removes dirt by opening up grimy cellulose areas while leaving cotton fiber alone. Another microbe makes an enzyme called "alkaline amylase," which will join glucose molecules into another molecule called "cyclodextrin"—used for slow-release capsules. Last but not least is the bacterium that will digest oil spills.

Then there's rock varnish. That's a dark coating less than half a millimeter thick, composed of minerals (mostly clay and manganese) now thought to be deposited by bacteria that live on porous rock surfaces that are at least intermittently wet. Some rock varnish is ancient and can be dated. Taking a bit of surface varnish over a pictograph makes it possible to date the underlying rock art without defacing it.

Rock varnish indicates what past environments were like and how climate changed. It thus identifies "stable land forms" so that people can live safely on parts of what geologists call "alluvial fans" without worrying about getting flooded someday. Surface stability is also shown, in case anybody wants to find a safe place for stashing away toxic wastes.

The large and very important diazotrophic cyanobacterium *Trichodesmium* is a phytoplankter. Besides having a fancy name, it seems to be the main fixer of nitrogen in the tropical North Atlantic Ocean. Plants need this nitrogen, and the ocean feeds many animals, including us.

The new microtechnology has come up with a tremendously important use of microbes—finding ways of growing them and bioengineering them to make products we need. Human hormones, drug receptors, and growth factors can be mass-produced by bacteria, now to be fed on food that's primarily processed algae. Even more important, microbes help make it possible for drugs to be targeted to the areas of the body where they are needed. For instance, certain bacteria produce enzymes that help them orient in magnetic fields. Using these enzymes in a drug, doctors can guide it to the diseased area by placing a magnet there.

Other microbes, genetically engineered, will deliver pesticides or fertilizers right to a legume plant (alfalfa, beans, clover, and peas). In the genetic code of the microbe is a "promoter" that responds like an on-off switch to the plant's chemical messages. Using gene-splicing techniques, scientists attach the on-off switch to another gene that will deliver the pesticide or the fertilizer only to the plant's roots. This means that lower concentrations of pesticides and fertilizers can be used, altering the environment less.

Microbes are also beautiful because they are our remotest ancestors, the precursors of all life on Earth. We and all other multicellular creatures are composed of eukaryotic cells. Many scientists now believe that all eukaryotic cells are composed of former free-living bacteria that learned to get along with and inside each other.

If each of us is a community of cells that are each a community of living things, is it too much to hope that some day all intelligent life will also be a cooperative community?

Delightful Diversity

Diversity really means variety. Human beings like variety. We used to call some of our shopping places "variety stores." In countries where consumer goods have been state-regulated, their citizens are now clamoring for the variety other people enjoy.

The trouble is that human beings, choosing which suit or car to buy, which TV program to watch, forget that diversity is what our whole planet is about, and that we humans have been "state-regulating" our poor planet into a sameness that is not only boring but dangerous.

The biggest part of the problem is that there are too many

humans. Too many of any one species is bad, but if that species is technologically equipped to use up and destroy its habitat, it's going to be in trouble. Being human means using technology, and that includes the primitive slash-and-burn technique for clearing land—because use of fire is technology, one of the earliest we acquired.

Countries without oil (which will run out anyway for everybody, sooner or later) are trying to diversify their energy sources. Other states and cities worry about diversifying work, especially when an area has been too heavily concentrated on one industry. When the Soviets lost interest in the nuclear-arms race, and then ceased being Soviets altogether, quivers of understandable fear went through American defense factories.

But it's the planet itself we have to worry about, for we are part of its life and totally dependent on it. When we humans everlastingly try to "control nature," we lose control of our place in nature and seem to be on the way to rendering ourselves homeless. Humans should diversify their habitats so some of them live in lunar or Martian domes, or in orbital space colonies, but since we haven't been willing to invest in this safety factor, we'd better take care of our one home.

Ecology, from the two Greek words for "house" and "science," shows that our "house," planet Earth, has an incredible diversity of life. New explorations of the rain forests have shown that the number of species on Earth has to be drastically revised upward. Are all these species necessary, especially to us? How about the pupfish, the panda, rare orchids? Necessary?

We can't tell. We may never be able to know exactly what's necessary for our own welfare. Therefore we'd better be very conservative in our treatment of the world's diversity, because we don't know what mistreatment will do to us. As fast as the rain forests go, scientists are discovering new medicines obtainable from plants threatened with extinction. Cattle and sheep will each denude a land that can otherwise support many different herbivorous creatures. Ugly alligators should live, for they make "gator holes" that become pools, and build up "gator nests" that make islands.

Everything's connected. The diversity of life on Earth is intricately interwoven, the parts overlapping, protecting each other's existence. Ecosystems work best when complex, with various plants and animals fitting many niches, doing many things, not necessarily competing.

For us, diversity is risk insurance. We put ourselves at terrible risk of disaster by concentrating on just a few food animals and plants whose genetic diversity is not enough for them to withstand major changes in the environment. We're told to eat more fish, but we haven't protected the waters where fish live, and we've overfished areas without helping to restock. We like and need many plants, but we don't protect the creatures that pollinate those plants—insects, birds, bats.

Imagine a scenario in which the planet is left with only human beings and whatever humans create. This is not feasible, probably because we'd never last that long, but just suppose. Chances are we'd reinvent diversity—millions of different kinds of robots to fill every niche. Then we'd argue about whether or not to control all of them, and either we'd decide not to, or we wouldn't be able to. Eventually there'd be robots quietly planting trees, pollinating plants, and patrolling for poachers.

Anyway, the truth is that we enjoy diversity, and treasure wildness, as Thoreau named it. We like to go to the ocean, the mountains, the desert—where we feel part of something larger than ourselves, a planet that will go on living after we've gone.

Besides, would you be satisfied with eating just one fruit? Or with conventional sliced white bread? Remember that diversity even tastes good!

II

OUR PLANET AND OUR NEIGHBORS

Mantle
and Core

Worried about the economy? Politics? The apparently inerad-
icable tendency for human beings to mess up not only their
own lives but the very planet they live on? Certain sciences
present such a wide, deep, or far perspective that it's a relief
to delve into them and forget the vicissitudes of ordinary life.

Astronomy and cosmology fit the bill, but sometimes they
scare people because they seem removed from humanity. Isaac
once had a phone call from a sobbing young person who asked,
"Is it really possible for the universe to end someday?" It's
possible, but not worth worrying about right now.

Then there's geology. Rock hounds collect their specimens
in what must be a pleasant escape, but geology is not all rocks.
It's about as close to home as any science can get, because
geology is about our home, Earth. Geologists are the scientists
who study things like volcanic eruptions (only too likely to
affect human living) and earthquakes (ditto).

Yet when geologists and geophysicists are studying the
deep structure of the Earth, the perspective is certainly broader
than most ordinary human problems. The planet's geophysical
history outranks all the romantic histories of all the countries
sitting on Earth's crust, only 64 kilometers (40 miles) thick.

Our solar system was formed about 4.6 billion years ago
from condensation of gas and dust clouds. As the gas and dust
ball grew, the light stuff collected outside as atmosphere while
the heavy stuff mashed together as the planet itself, pressure
and heat increasing deep inside. This inside of Earth is a re-

markable place, although it bears no resemblance to Jules Verne's description in *A Journey to the Center of the Earth*.

Under Earth's rocky crust is the mantle, about 2,900 kilometers (1,800 miles) thick. The mantle is rocky too, but the rocks are olivine type, rich in magnesium and iron, denser than crustal rock. You wouldn't think there'd be any water in the mantle, but there is, structurally bound to the minerals as hydroxyl. The mantle's rocks are now thought to have as much water, in the form of hydroxyl, as a good percentage of the oceans and seem to be an important part of the recycling of water through the top layers of Earth. Recent studies also indicate that the hydrogen in the hydroxyl may contribute to the mantle's electrical conductivity.

Beneath the mantle is the planet's core, under pressure ranging from 10,000 tons per square inch at the top to 25,000 tons at the center. After studying seismographs, British geologist Richard Dixon Oldham established in 1906 that Earth's rocky mantle ended at a liquid region. For years it was believed that the entire core was liquid, but now it's known that there's an outer core of molten nickel and iron, six times wider than the solid inner core that floats in it like a tiny planet 5,000 kilometers beneath our feet.

As Earth cooled down in its early development, the inner core formed by slow solidification of the liquid core. It may have taken as much as 3.6 billion years for the solid core to reach its present size. Canadian geophysicist Dr. Douglas Smylie recently studied and demonstrated the way the solid inner core oscillates in the liquid outer core. Eventually, the new techniques may refine estimates of the inner core's mass.

The two aspects of Earth's center form a "geodynamo" powered by the heat convection of the outer core's fluids plus the growth of the solid inner core, now so big that the gravitational energy it contributes may be the most important part of the geodynamo. Geophysicists at Cambridge University have come up with an analytical model, based on global heat conservation, that explains (to other geophysicists, not to me) the developmental history of the core.

It may seem as if this picture of Earth's insides—slowly sloshing molten metal around a very slowly growing solid ball—is too out of sight and, for most of us, too out of mind to be important, but it is. This inner geodynamo generates Earth's magnetic field, which diverts some of the lethal cosmic radiation bombarding the planet. The periodic reversals of Earth's magnetic field may be channeled by alterations, not just in the fluid outer core but in the overlying rocky mantle itself. Research continues on the enigma of magnetic reversals.

The boundaries between core and mantle and between mantle and crust are under intense study. In certain persistent "hot spots" starting near the boundary of mantle and core, mantle material rises up like a plume to break through the crust as volcanoes. Islands like the Hawaiian chain are thought to be formed as the tectonic plates of the crust move over mantle hot spots. Roger Larson has hypothesized that a "superplume" erupted from the mantle 120 million years ago and caused not only crust formation but a temporary (well, 40 million years' worth) stabilization of Earth's magnetic field polarity. Superplumes may increase the rate of movement in the liquid outer core, altering the force of the geodynamo.

Beneath the fragile layer of life clinging to Earth's crust are the mantle and core, the heart of our planet and well worth studying.

The Oldest Rocks

Geologists, for many years, have tried to find really old rocks on Earth's surface, from which they might deduce Earth's earliest history. It's not easy, for you must find rocks that solidified billions of years ago and have rested in the ground ever since without being seriously disturbed.

A few years ago, Samuel A. Bowring, of Washington University in St. Louis, and his colleagues discovered rocks in northwestern Canada that seem to be 3.96 billion years old. They formed, apparently, when the Earth was only about 600 million years old, or only one-eighth its present age.

How is it possible to know that the rocks are that old? The answer seems to lie in tiny crystals of zircon that exist within the rocks. Zircon is "zirconium silicate"; that is, it is a rocky substance that contains atoms of the not very rare metal zirconium, together with atoms of silicon and oxygen.

When the zircon crystals form, they set up regular lattices of atoms of zirconium, silicon, and oxygen. There are, of course, other types of metallic atoms in the vicinity. Some of these metallic atoms can fit into the lattice and can replace an occasional zirconium atom. Other metallic atoms cannot fit and therefore stay out of the tiny crystal.

The fortunate thing about the zircon crystal is that it can accommodate uranium atoms that might be in the vicinity, but not lead. The result is that the zircon crystals have tiny amounts of uranium present, but no lead.

At least, they have no lead at the start, but they *develop* lead because uranium atoms are radioactive. One of them occasionally breaks down and forms another type of radioactive atom, which also breaks down to still another type, and so on. Eventually, though, the breakdowns end in the formation of a lead atom. The lead atom is stable and remains.

The breakdown of uranium is not very rapid. It proceeds so slowly, in fact, that it takes fully 4.5 billion years for half the uranium in a zircon crystal to turn into lead. On the other hand, the breakdown is very regular and follows simple rules that have been worked out accurately in the laboratory. If a zircon crystal is analyzed and found to contain so much uranium and so much lead, one can calculate how long it must have taken the uranium to break down and produce that lead, and that, in turn, tells you how old the rock is.

Of course, things aren't quite as simple as this sounds. Making the actual measurements and interpreting them properly is not necessarily easy. The logical thing to do might be to take the entire zircon crystal and analyze it for uranium and lead content. Unfortunately, nothing is perfect. The zircon crystal may have tiny hairline fractures in it through which lead may have leaked out.

What one must do is to analyze different parts of the tiny crystal to find those parts in which the lead content is a maximum and where the lead has consequently suffered the least loss.

In order to do this, Bowring took his rocks to Australia, where they had a machine that was just right for this sort of measurement. It fired a beam of charged particles at the zircon crystal, and the energy of impact vaporized about two-billionths of a gram of material. This tiny bit of zircon vapor was then analyzed by means of something called a "mass spectrometer," which counts the lead, practically atom by atom. And thus the rocks were found to be 3.96 billion years old.

Oddly enough, still older zircon crystals have been found. Tiny crystals of zircon found in Australian rocks have measured out to be 4.3 billion years old. However, these crystals are located inside relatively young rocks. They must once have been part of ultra-old rocks, but the forces of erosion broke down those rocks and the crystals were then incorporated into newer rocks. The mere existence of these ultra-old zircon crystals doesn't tell us anything about the early Earth. We need to find such ultra-old crystals in their original rock, and there is no way of knowing whether geologists will ever succeed in doing so.

Meanwhile, the rocks of northwestern Canada are interesting. They are granitic in nature—the kind of rocks that make up the continents of the Earth. This would indicate that nearly 4 billion years ago, continents already existed on Earth.

What's more, these granitic rocks are not what we would expect of primeval rocks. Everything geologists have learned makes it clear that such granitic rocks have evolved from simpler predecessors. This means that already at 4 billion years ago, Earth had undergone complex changes from the moment of its formation.

Older Than
We Thought

Every once in a while, as new information comes to hand, scientists must make changes in theories in which, until then, they had felt quite confident. This happened when drillings into coral reefs forced revision of some dates that archeologists had been relying on.

The dates in question are those obtained by studying the carbon 14 content of old objects. Carbon atoms come in three varieties, carbon 12, carbon 13, and (in tiny traces) carbon 14. Of these, carbon 12 and carbon 13 are stable, but carbon 14 breaks down slowly. As long as any organism is alive, it continues to incorporate fresh carbon into its tissues, including carbon 14. The new carbon 14 is incorporated as fast as the old breaks down, so the quantity within the tissue remains steady.

After any organism dies, however, the carbon 14 it contains breaks down, but no new carbon 14 is incorporated into the dead material. This means that the quantity of carbon 14 slowly declines, and from the amount by which it declines, scientists can tell just how long it has been since the material was alive.

In this way, scientists can date old pieces of wood or charcoal and tell how long it has been since they were part of living trees. They can date old seeds or bits of dead coral and tell how long it has been since they were alive.

The method looks perfectly reliable. Carbon 14 breaks down at a steady and unchanging rate, and measurements of the quantity present are delicate and precise. What can go wrong?

A lot depends on how much carbon 14 (in the form of carbon dioxide) there is in the environment to begin with. If the quantity rises, more is incorporated, and if the quantity falls, less is incorporated. Volcanic action increases carbon dioxide content in the air, and so does an increase in cosmic ray activity. A fall in ocean temperature increases the amount of carbon dioxide that dissolves in the ocean, and all these effects can operate in reverse, too.

This means that scientists can't be certain how much carbon 14 "ought" to have been in the old materials when they were alive, and therefore how old the current level indicates it to be. The simplest assumption is to suppose the carbon 14 level in the environment has always been pretty much the same, but it isn't really safe to do so.

The best thing to do is to correlate the carbon 14 findings with other ways of checking age. For instance, some ages can be determined both by tree rings and by carbon 14. Tree-ring data indicate the changing levels of carbon 14, and there was reason to think that carbon 14 ages, as usually determined, were a little too low, but it was hard to tell by just how much, especially since tree-ring data only goes back 10,000 years.

The drillings in the corals give ages much older than that, and they can be checked by determining the quantity of uranium and thorium isotopes that are present. They also break down at a steady and known rate, but there is no question of changing quantities of those isotopes with time so that the uranium/thorium ages are more reliable than the carbon 14 ages, especially now that the measurement techniques of the uranium/thorium isotopes have been refined and made more accurate.

What it shows now is that there is a discrepancy of about 20 percent. Carbon 14 data seemed to show that the peak of the most recent ice age was 18,000 years ago, but it now seems that actually it was 21,500 years ago.

Whenever such changes are found necessary, nonscientists often say, more or less gleefully, "All the textbooks will have to be rewritten," but that is rarely the case. Actually, it just means a few paragraphs will have to be modified. The new attitude toward carbon 14 ages doesn't alter the order in which things are supposed to have happened. It just makes everything that has been subjected to carbon 14 dating up to 20 percent older than had been thought.

This is not to say that such changes can't be important in other ways. For instance, many scientists think that the ice ages of the last million years came and went in accordance with slight cyclic changes in Earth's orbit that meant the delivery of slightly more, or slightly less, heat to the Earth.

Those cyclic changes, if calculated correctly, indicated that a slow heating trend began about 23,000 years ago. This lowered the rate at which the glaciers were accumulating and, finally, after a few thousand years brought them to a halt and forced them to begin retreating. A wait of 5,000 years before the peak came and the retreat began was a little too long, and scientists wondered why the glaciers persisted. But now, if the peak was 21,500 years ago, that peak came only 1,500 years after the heating trend began, and that is much better. For this theory, at least, the modified age determination system is welcome indeed.

Water—
the Circulation Below

Earth has an ocean. In spite of the different names for different sections, the ocean is one entity, proven by any photograph taken from space. Our planet is a beautiful blue because it has an open, liquid ocean, unique in the solar system. Europa, a satellite of Jupiter, may have liquid water under its frozen surface, but nothing rivals Earth's ocean.

Even in ice ages, liquid water covers over 70 percent of Earth to an average depth of 3,730 meters. This ocean fosters life, even on land, for water evaporates from it, enters the circulation above, in the atmosphere, and comes down on land as rain or snow.

The ocean moves, not just as tides back and forth upon the land, but in great currents of flowing water. As the Earth spins, its surface rotates at over 1,000 miles per hour on the equator, moving slower north and south to the poles. In 1835 French mathematician Gaspard-Gustave de Coriolis studied the effect of Earth's rotation and showed that it caused what has been named the "Coriolis effect"—any moving body on or near Earth's surface appears to move sideways.

The Coriolis effect applies to the ocean, in which currents move clockwise in the Northern Hemisphere and counterclockwise in the Southern. Many currents are famous, especially the Gulf Stream, which the American Matthew Fontaine Maury called "a river in the ocean," the largest river on Earth. New data show that this is not an accurate description, for although the Gulf Stream looks like a single, separate entity in the Atlantic, it is actually a system of currents carrying warmer water into colder sections.

In spite of the pull of gravity, the ocean is not "level"—there are upgrades and downgrades, caused by the Coriolis effect. Scientific satellites sent up in recent years show that thirty degrees north or south latitude, there's a distinct bulge in the ocean between the prevailing winds. There's also an actual slope of 4.7 feet from the outer edge of the Gulf Stream system to the inside!

There are two kinds of circulation in the ocean. One is largely surface, driven by the wind. The other is the deeper and slower thermohaline circulation, caused by differences in water density and temperature. Saltier or colder seawater sinks and moves under surface water.

As the ocean's water circulates, it moderates the overall temperature of the planet, not only by dispersing heat but by absorbing gases that would otherwise add to the greenhouse effect we humans seem determined to foster by burning fossil fuels. We need our ocean, and we need to keep it clean and healthy. The urgency of this has prompted intensive study of the ocean.

Australian oceanographers Nathaniel Bindoff and John Church have determined that deep water in the Pacific has increased in temperature .01°C since 1967. That doesn't sound like much, but such warming means that seawater has expanded and sea level risen almost three centimeters so far.

Because the circulation in the ocean is so important, scientists study not only its present patterns but those in the past. It has been thought that glaciation may be helped along by fresh meltwater (from glaciers) that decreases the ocean's salt content, wiping out density differences between deep and surface water. Then there's nothing to power the movement of surface to deep water and back. This "circulation loop" slows, and warm tropical water doesn't go north as it ordinarily does. The theory is that this chain of events produces an ice age, which in itself ties up fresh water in ice, so there's no more meltwater, which eventually produces density differences.

Then circulation begins, warm water can proceed north again, and the ice age ends.

Paleo-oceanographers E. Jansen and T. Veum disagree with this theory of ocean circulation. Studying microfossil carbon isotope data, they believe that the deep-water circulation continues even in ice ages, although it somehow doesn't enable warm water to flow north. More studies are under way to determine whether or not the deep—thermohaline—circulation does or does not shut on and off, or continues to operate at different levels.

It's said that "he who knows no history is doomed to repeat it." I think that he who doesn't understand Earth's past will be unprepared for its future. If we're giving Earth a greenhouse situation, we'd better find out how the ocean's circulation has operated in the past and is likely to operate now.

The latest news about ocean pollution is grimly depressing. Overfishing. Coastlines clogged with sewage. Birds and animals dying of toxicity, disease, or from being trapped in plastics and the worst kind of fishnets. Coral reefs plundered and killed. Is there any hope?

A little. More articles, more people interested. There's a new computerized oceanographic probe that can monitor water quality and tell the monitoring base about amount and location of pollution. Scientists are also studying the flatfish *Limanda limanda,* a bottom-dweller whose liver reveals molecular and chemical changes due to pollution. Right now those livers are showing fatty degeneration, a cancer precursor.

Humanity should grow milkweed. Not only does it attract monarch butterflies, but its fibers (along with cotton and kenaf) are good at soaking up oil slicks on the ocean. Every little bit helps. So would changing our habits.

Air—
the Circulation Above

To a physician, the word *circulation* brings to mind a diagram of human blood and lymph vessels neatly drawn in red and blue. To others, *circulation* might mean the number of readers for a magazine, or how efficiently a freezer's motor moves its cold air into the rest of the refrigerator.

More important is the circulation of our air, for the oxygen-rich atmosphere of Earth exists nowhere else in the solar system. Our life-sustaining atmosphere is roughly 21 percent molecular oxygen and 78 percent molecular nitrogen, with about 1 percent argon, plus trace constituents like carbon dioxide, helium, neon, krypton, xenon, nitrous oxide, methane, and carbon monoxide.

In the high upper levels of the atmosphere, where the 1:4 mixture of oxygen and nitrogen is much thinner, high-energy radiation from the sun breaks down some oxygen and less nitrogen to the atomic form. Somewhat lower down, the sun combines atomic oxygen with ordinary molecular oxygen to make a three-atom variety called "ozone"—that word appearing so often nowadays in newspapers and scientific journals. Ozone is unstable, so there's never a lot of it, but what there is shields Earth from the sun's lethal ultraviolet radiation.

All well and good, except for circulation. The atmosphere of Earth doesn't just lie like a blanket around the planet. It moves, and stuff moves up and around in it. The atmosphere has been likened to an ocean of air with us living at the bottom of it. What we do to our part of the air ocean affects the rest, thanks to circulation.

Earth is round (well, a lumpy oblate spheroid) and spins as it orbits the star we call "Sol." The atmosphere moves too, thanks to the spin of Earth plus the sun's heat, absorbed by and reflected from the planet's surface. Since hotter air rises, tropical air goes up and out north and south to Earth's poles, then cools, sinks down, and is whipped into wind by Earth's rotation. In the Northern Hemisphere, the atmosphere flows in a skewed clockwise direction (vice versa below the equator). We call much of what happens in this circulation "weather."

So far, human beings can't control weather, but perhaps Earth—as a living planet—partly regulates its atmosphere with biological means. A recent study of one greenhouse-increasing gas, methane, shows that some soil bacteria can oxidize it even in deserts. Bacteria in forest soils do the best job, but humans are destroying the forests while adding to the rise in methane by burning fossil fuels.

In 1992, 150 nations tried to set strong international limits on human production of gases contributing to the greenhouse effect. Because the United States refused to agree, no strong limits were set. At the same time, two headlines appeared, in different publications, on the same day. One was OZONE LAYER SUCCUMBS TO ASSAULT and the other was OZONE SURVIVES. The first article showed that chemical pollutants destroy more of the ozone layer than was previously thought. The second article said that the hole in the ozone layer over the Arctic was not as damaged as expected. If you read past the headlines, it's clear that both articles agree—in the future things will get worse. The circulation of our atmosphere can't cope with the pollution we're putting into it, and the ozone layer will diminish.

There are other planets with atmospheres. Venus has one that also circulates with the planet's rotation. Data on Venus amassed by the Galileo and Pioneer Venus probes are being studied carefully, especially by astronomers M. D. Smith, P. J. Gierasch, and P. J. Schinder. Apparently, the thick clouds of the Venusian atmosphere form near the equator, move toward the poles, and are sheared as wind into a characteristically streaky, Y-shaped pattern moving around Venus every four or five days. This pattern, not visible except in ultraviolet images, is thought to be a special kind of wave called "Kelvin"—powered by gravity, altered by the effects of the planet's rotation, and probably maintained by the "cloud feedback" of heat. This is interesting, but the fact remains that the mainly carbon dioxide atmosphere of Venus is trapped in an overpowering greenhouse situation, much too hot and poisonous for humans to breathe.

Then there's Mars—no Martians, probably no life at all, and an atmosphere that's very thin and oxygen-poor.

Back to Earth. Recently, one of my phone conversations with midwestern friends was interrupted by a loud noise at their end. They told me it was just the local tornado alert. It came to nothing, but for a while it brought home to me how the circulation of Earth's atmosphere both sustains and endangers human beings (except for lucky Dorothy whizzing off to Oz).

We can curse the vagaries of our weather, the polluted air we ourselves have produced, the dangers of a potential greenhouse effect, and the waning of our protective ozone layer. But as the data on other planets show, we can't go anywhere else, not yet. Yet we can do two things: (1) take care of our own planet, and (2)—in case we don't succeed at number 1—build protective habitats for ourselves elsewhere, on the Moon or Mars, or in rotating artificial worlds we devise. We'd better get started.

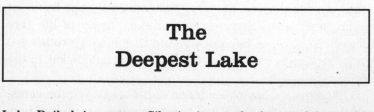

The
Deepest Lake

Lake Baikal, in eastern Siberia, is not the largest lake in the world. That honor belongs to Lake Superior, which is 31,500 square miles in area, whereas Lake Baikal has an area of only 11,780 square miles. However, Lake Superior has its greatest depth at 1,333 feet, whereas Lake Baikal has a depth of 5,713 feet, better than a mile.

Lake Baikal is the only lake that has deep-sea fish in it. Because of its depth, Lake Baikal contains one-fifth of the entire supply of Earth's fresh water. In fact, it has twice as much fresh water as all five of the Great Lakes put together. It is important, then, to find out just how Lake Baikal works. For instance, how does it manage to circulate its water so that the deepest layers possess plenty of oxygen and can support life?

Recently, Dr. Ray F. Weiss of the Scripps Institution of Oceanography made use, for the purpose, of the gas Freon, usually used in refrigeration devices and in aerosols, which has now been recognized as a gas that is helping to destroy Earth's ozone layer. Because Freon is very stable and undergoes no changes, it can be used as a marker. It can be added to the water of Lake Baikal, and its presence here and there in the water will tell us how it circulates.

Weiss discovered that 12.5 percent of the deep water in Lake Baikal is renewed each year, so that every eight years or so, the water in the lake's depths has a fresh supply of oxygen.

An even more important discovery was made in Lake Baikal in the mid-1980s by a team led by Kathleen Crane of the Department of Geology and Geography of Hunter College.

In recent years, it has been discovered that the oceans have places where there are "hot spots" or "hydrothermal vents." In such places, there are materials that arise from the depths, and on such materials, there are forms of life that flourish. Now it has been discovered that such hydrothermal vents are also to be found at the bottom of Lake Baikal, in the northeast corner.

Photographs have been taken of the areas near the vents, and they show that there is a nearly continuous layer of bacterial life there, consisting of long, thick white strands inside a matrix of whitish material. These hot spots are hot all right. Whereas the water has a temperature of about 3.5°C, the material under the bacterial mat has a temperature of 16°C.

Nor is the life that exists near the vents purely bacterial in nature. There are white sponges and other animals called "gastropods" and "amphipods." They are not found in parts of the lake bottom that are removed from the hot spot.

The animals found in Lake Baikal seem to have certain resemblances to forms found in saltwater, so it may be that the lake was once connected to the oceans. On the other hand, it may be that Lake Baikal is spreading and that it is the nucleus of a new ocean that may appear someday. We don't really know.

It's a little puzzling as to exactly where the hot spots at the bottom of Lake Baikal appear. In the ocean, hot spots appear where the sea floor is spreading, where melted rock is coming up from the depths. This is not so in the case of Lake Baikal.

The hydrothermal vents in Baikal are found along a fault that is more than eighteen kilometers from the axis of the rift valley floor. It is suspected that this depends on places where magma, or melted rock, may exist and may be extruded upward. It may be that Lake Baikal has comparatively little of this magma and hence not much in the way of hot spots.

Lake Baikal, by the way, is an isolated life-community and has a large number of plants and animals that are peculiar to itself. There are over twelve hundred different species that live at different levels in the water and, of these, some three-fourths are not found anywhere else.

Of the fifty species of fish in the waters, salmon and whitefish are the most heavily fished. The largest of the Lake Baikal fish are sturgeons, some of which measure 71 inches and weigh as much as 265 pounds.

There is only one mammal native to Lake Baikal, and that is the Baikal seal.

Measures are being taken to prevent the pollution of Lake Baikal. After all, it is unique—there is no place like it anywhere else on Earth. It is important to keep it as pristine as possible, and to protect the plants and animals that belong to it.

The
Big Melt

Hurricanes do a great deal of damage, but they are pinpricks compared to some of the natural disasters that may have struck the Earth. This does not refer to the cometary strike, 65 million years ago, that killed the dinosaurs, but to possible events that took place only a few thousand years ago, when human beings were about to begin establishing civilizations in the Middle East.

For the last million years, Earth has had intermittent periods in which huge ice sheets covered the northern half of North America and large parts of northern Eurasia as well. These may have been caused by slight, periodic changes in Earth's orbit, and the results have been serious in the last million years only because the movements of Earth's crust have surrounded the North Pole with land areas.

Apparently, there are periods when the summers in the Northern Hemisphere cool off a bit. In that case, not all of the winter snows have a chance to melt before the next winter snows begin. Little by little, from year to year, the snow cover increases. As it does, more and more sunlight is reflected into space and does not have the chance to warm the Earth because snow is more reflective than bare ground is. Summers therefore grow cooler still.

Thus, the glaciers slowly formed and moved southward,

reaching as far south, at an extreme, as the Ohio River and Long Island. The sea level dropped and land bridges connected Asia with North America in the north and with Australia in the south, so that human beings made their way into those continents from their Old World origins.

But then, the summers warmed a bit as Earth's orbit cycled back in time, so that more snow was melted than fell in the next winter. Less sunlight was reflected and more was absorbed, as bare ground was uncovered so the summers grew warmer still. Little by little, then, the glaciers retreated. Ten thousand years ago they were completing their most recent retreat, and the world became as it is now.

Usually, though, the coming and going of the ice sheets is thought of as slow—glacially slow, in fact. There is no way of imagining the coming of the glaciers as anything but slow, but what about the melting?

Back in 1975, Cesare Emiliani of the University of Miami studied the fossil remnants of microscopic organisms under the sediments on the floor of the Gulf of Mexico. From his studies, he concluded that there was a period, eleven thousand years ago, when the Gulf of Mexico contained water that was much less salty than it is today. He suggested that the ice sheets had undergone a sudden melting and that a vast flood of water had entered the Gulf of Mexico and raised the sea level markedly.

The suggestion was largely ignored because it was difficult to imagine the ice melting that fast, but in 1989, John Shaw of Queen's University in Kingston, Ontario, made a suggestion as to just how such floods might come about.

The regions where once the ice sheets were found have a scattering of low hills called "drumlins." These are usually supposed to have been formed by the grinding action of glaciers as they came and went. Shaw, however, feels they may more easily have been formed by a vast rush of water.

He suggests that the ice sheets did indeed melt very slowly, but that the water did not necessarily run off, soak into the ground, pour into rivers, and reach the sea as rapidly as it formed.

Instead, water might have slowly settled down to the bottom of the ice sheet, soaked into the ground till it reached bedrock, and then slowly accumulated there. There would thus form what was essentially a lake of water underneath the ice sheet, and this would be prevented by ice dams from spreading outward.

Eventually, though, as the glaciers continued very slowly to melt, sections of the ice dams would weaken and then break. The lake of ice water that had been pent up would then pour out seaward in a vast flood that beggars anything we can imagine.

Shaw has calculated that something like twenty thousand cubic miles of water may have poured out of the ice, all at once, to form the drumlin fields of northern Saskatchewan. The Amazon River, the largest on Earth, takes ten years to discharge twenty thousand cubic miles of water into the Atlantic Ocean, but the ice lake may have discharged it in a matter of a few days only. It would therefore have the effect of a river perhaps a thousand times as large as the Amazon.

That water, tumbling into the ocean, may have raised the global sea level as much as nine inches in just a few days. The rising water would have moved up the low-lying continental shelves that had been exposed during glaciation. Human beings, retreating inland before the inflowing water, may have reminisced and exaggerated afterward, giving rise to tales of drowned continents and universal floods.

Moon Rocks

In 1990, a very unusual rock from Antarctica was discovered to be a fragment of the Moon. People wondered how a piece of the Moon could be in Antarctica, but that is not as mysterious as it may sound. The Moon and all the bodies in the solar system were formed by the coming together of smaller fragments. Once the planets and satellites approached their present conditions, something over 4 billion years ago, there were still final bodies slamming into them. Indeed, at a much reduced rate, those collisions are taking place even today.

We see the signs of those collisions in the craters that exist on so many worlds that don't have atmospheres or oceans or lava flows to wipe them out. On Earth, the signs have mostly been erased, but on the Moon they exist untouched, and our satellite is covered with craters.

Each crater is the result of a meteor, sometimes quite a large one, that slams into the Moon at a speed of twenty miles a second or so. Such an object colliding at such a speed produces a huge explosion in the Moon's surface, driving material upward.

The same has happened on Earth, but Earth's gravity makes it necessary for material to be driven away from the surface at a speed of seven miles a second in order for it to get away. Even a large meteorite strike won't produce such speeds, so exploded material falls back to Earth. The Moon is a smaller body with a smaller gravitational pull. Objects need to move at a speed of only 1.5 miles per second to escape. The result is that the meteoric bombardment of the Moon drives small pieces of material permanently away from the Moon.

The Moon pieces are not remarkable. The largest are probably very little more than gravel-size, and most of it is dust.

Some of it gets swept away by the solar wind that pushes it out into the farther reaches of the solar system. Part may eventually fall back to the Moon. Part, however, remains, and the space between the Earth and Moon is a little dustier than regions in far outer space would be. And every once in a while one of these Moon pieces loops out far enough to collide with Earth.

The Earth is constantly being bombarded with tiny meteorites, few of which are large enough to survive passage through the atmosphere and reach the planet's surface. Most of these meteorites are "primordial"; that is, they have existed in space ever since the solar system was in the process of formation. Others are the remnants of dead comets. But just a few are Moon pieces.

How do we study meteorites? Some are easily recognizable because they are chunks of metallic iron, which do not occur naturally on Earth. At least 90 percent, however, are rocky objects that cannot easily be distinguished from Earth's own rocks. Unless such rocky meteorites are actually seen to fall, they can't be located without a great deal of difficulty, and even if they are, they may, with time, have become contaminated with Earth's own material.

There is one unusual exception to this. Surrounding the South Pole is the continent of Antarctica, which is 5 million square miles in area and is covered by a thick, unbroken layer of ice. In recent years, Antarctic explorers have been finding occasional rocks on the ice surface. Any rock in Antarctica that has not been brought there by human beings must be a meteorite. There is no other way in which a rock can be lying on the ice. As a result, it has become possible to study meteorites with greater intensity than ever before in history.

But given that, we now have the second question: How can you tell that a particular meteorite originated on the Moon? That comes with chemical analysis. The Earth and the Moon are composed of the same chemical elements, but they are present in different proportions because the two bodies are of different sizes and have had different histories. In a sense, the proportion of the different elements represents a kind of "fingerprint" of the planet. Thus, primordial meteorites, even when rocky, are, for instance, considerably richer in iron than either Earth or Moon rocks would be.

Nearly a dozen meteorites in Antarctica have now been found to possess an elementary makeup exactly like that which is found on the Moon, and the conclusion is that such meteorites are pieces of the Moon. The first Moon rock on Antarctica was found in 1979, and it weighed about two ounces.

The largest Moon piece so far found was reported in 1990 by Jeremy Delaney of Rutgers University. It weighed about one and a half pounds and was a little over two inches across. All the Moon pieces so far found come to nearly five pounds in weight, and they can be studied at leisure without having to go to the Moon for them.

In that case, why have we bothered going to the Moon? Because only in so doing, and in collecting Moon rocks on the spot, could we determine the chemical "fingerprint" of Moon rocks and therefore identify objects on Earth as having definitely come from the Moon.

New Questions about the Planets

Until this present generation, astronomers were gloomily certain that we could *never* know anything about Venus's surface, because the planet was encircled by a thick and unbroken cloud layer through which sight could not penetrate.

Radar, however, uses waves that are very much like light

waves but a million times longer. The advantage of radar is that it can penetrate clouds, mist, and dust. It can go right through Venus's cloud layer as though it weren't there, strike the solid surface beneath, and be reflected. The reflected waves pass through the clouds again and can then be detected.

The disadvantage of radar is twofold. First, we didn't have any biological way of detecting it, no "radar-eyes," so to speak; and second, the long waves don't "see" as clearly as the tiny waves of light do. Both disadvantages have been corrected. We now have instruments to detect the reflected radar waves and devices for sharpening the "vision."

The result is a great improvement on previous attempts for radar-mapping of Venus, and we can see the surface as clearly as we can see that of the Moon. Naturally, some of the results are surprising. As in science generally, new observations always bring new questions.

For instance, several large craters have been detected on Venus but almost no small ones. This, in itself, can be explained. Venus's atmosphere is nearly a hundred times as dense as Earth's so that small meteorites are heated and vaporized away more efficiently than by Earth's thinner air. Only large meteorites can survive the trip through Venus's atmosphere, and it is they that cause the large craters.

What is puzzling, though, is that the dust and debris cast out of the large craters form a broken circle. They resemble the somewhat separated petals of a flower. This sort of pattern is found on Mars, and it is usually explained by the action of water. When there is no water on a world, as in the case of the Moon, for instance, the circle of debris forms an unbroken circle.

Now, here is the catch. Venus is completely and totally dry, so why are the craters surrounded by petallike structures? Can it be the wind (there is no wind on the Moon) or the effect of gases forced out from below the surface when the meteorite hits? One suggestion I have not seen made, but which makes sense to me, is that the meteorite may penetrate far enough to throw up "magma," or liquid rock, from the interior. Liquid rock may have the same effect as liquid water.

Two billion miles from Venus is the planet Neptune, which has a fairly large satellite (somewhat smaller than our Moon) called Triton, photographed at close range by Voyager 2.

Photographs showed dark smudges on Triton's generally bright surface, frosted with frozen methane and nitrogen. It is as though someone had stroked chocolate-stained fingers over the globe. After studying the photos, astronomers boiled down possible explanations to the two they consider most likely.

One possibility is that the nitrogen surface, largely clear and whitish, may have regions where there are darker particles which absorb the weak light of the far distant Sun and grow warm enough to vaporize the solid nitrogen. There could be a sudden burst as a quantity of the solid turns to gas (a "nitrogen volcano"), blowing the dark material upward. Triton has a thin atmosphere, and its wind may then blow the material into a long oval patch, and these are the dark smudges.

Another possibility is that there are occasional small eddies in Triton's atmosphere, something like the "dust devils" with which we are familiar on Earth. These can cause the dark dust to swirl and settle it out in elongated fashion.

In this way, learning more than we previously knew about something may answer questions we had earlier, but it will also teach us entirely new questions to ask.

Thus, Uranus's satellite, Umbriel, is almost completely dark, but it has a white doughnut-shaped region. We don't know what it is. Uranus's satellite, Miranda, seems to have been broken apart and come together again so that it is a queer jumble of structures, including marks that look like a sergeant's chevrons. We don't really know what that's all about. Saturn's largest satellite, Titan, has clouds as opaque as Venus's. We badly need radar information about its surface.

And the planet Neptune is a tissue of puzzles, with extraordinary rapid winds and a large "dark spot." New questions everywhere!

The Atmosphere of Mercury

Mercury's atmosphere? That seems strange because it is well-known that Mercury doesn't have a "real" atmosphere. Mercury is very close to the Sun and very hot. It is also quite small and has a low gravitational pull. Hot gases are harder to hold than cool gases are, and Mercury's weak pull isn't much of a holder anyway. Therefore—no atmosphere.

It depends, though, on what you call an atmosphere. Our Moon, for instance, has no atmosphere in an Earthly sense. There is a vacuum immediately above its surface. However, there are considerably more gas molecules in a cubic foot of space near the Moon's surface than in a cubic foot of space far away from any planet. The Moon, therefore, might be viewed as having a very thin atmosphere, one that is only about a billionth as dense as Earth's. That's not much, but it's something and it's there.

In the same way, there is a very thin layer of gas in the immediate vicinity of Mercury's surface. The two elements in that gas are easily detectable. Sodium and potassium are metallic elements that melt into liquids at comparatively low temperatures. Mercury is not hot enough to boil these liquids, but it is hot enough to keep some of it in vapor form. (Thus, Earth is not hot enough to boil water, but it is hot enough to keep some water in vapor form in the atmosphere.)

Mercury can't hold onto such vapors, however. Any sodium or potassium vapor that existed in the past should have disappeared long ago. Since the vapors are still there, they must be manufactured, somehow, at the same rate that they disappear.

One possibility is that small meteors are constantly hitting Mercury's surface and that these bring new supplies of sodium and potassium, which heat into vapor. Another is that charged particles from the Sun (the "solar wind") slam into the surface of Mercury and knock sodium and potassium out of its rocks.

As it happens, though, there is a huge crater system on Mercury called "Caloris" (from the Latin word for "hot," because it faces the Sun when Mercury is closest to that body). Caloris was undoubtedly formed by a colossal meteoric slam in the early days of the solar system. It must have cracked and fractured Mercury's crust, and the crust has probably remained cracked and fractured ever since, because Mercury is likely geologically dead and so its crust remains in whatever form it has been pushed into.

Ann L. Sprague of the University of Arizona found that when Caloris is in view from Earth, about ten times as much potassium can be detected above Mercury's surface than when Caloris is out of view. The natural conclusion is that although any sodium or potassium on Mercury's surface would have been lost long ago, there are still large quantities of it below the surface. The subsurface supply is heated by the Sun, and small quantities leak upward through cracks and fissures in Mercury's crust. It would do so most readily where the surface was really smashed, as at Caloris, and that is why more potassium is detected when that feature is in view.

Studies such as this may give us better ideas concerning the interiors of worlds—interiors that we can't study directly.

For instance, the Moon also has sodium and potassium atoms present in its very thin atmosphere. Sodium has a smaller atom than potassium has and, in the universe generally, smaller atoms are more common than larger ones. There is, therefore, more sodium than potassium in the Moon's atmosphere. In fact, there is five times as much sodium as potassium in the Moon's atmosphere.

It is not surprising, then, that there is more sodium than

potassium in Mercury's atmosphere, too. However, sodium is *fifteen* times as common as potassium in Mercury's atmosphere.

One explanation that offers itself at once is that potassium is more easily converted into vapor than sodium is, because potassium has a lower boiling point. This means that on Mercury, which is much hotter than the Moon, subsurface supplies of sodium and potassium have been vaporized at a greater rate than on the Moon, and this is especially true of potassium. For that reason, the layers beneath Mercury's crust must be much more nearly stripped of their potassium than is true for the Moon. As a result, it is not so much that Mercury has more sodium in its interior, but that it has less potassium.

Probably all planets give off gases. Earth is geologically alive so it has volcanoes spewing molten rock and water vapor. Jupiter's satellite, Io, is heated by tidal action and has volcanoes spewing sulfur. Frigid Triton, Neptune's satellite, spews ice. Since the leakage of vapors from a planet's interior can be viewed as a very slow form of volcanic discharge, any sizable world may do that.

More on Venus

October 8, 1992, marked the death of one of Earth's most hardworking and faithful servants—the Venus orbiter called Pioneer 12. It entered orbit around Venus on December 4, 1978, and had been transmitting data ever since.

Pioneer 12, however, was only one of about thirty-five space probes investigating Venus since 1961, when the Soviet Venera I arrived at the planet. The first spacecraft to transmit internal data about the Venusian atmosphere was Venera 4, which broadcast for ninety-three minutes after entering the Venusian clouds on October 18, 1967. Since 1989, the Magellan spacecraft has been radar-mapping Venus, and it's hoped that it will continue to do so. This year there have been many scientific articles about the Venus data.

Venus is the second planet out from the Sun, and only slightly smaller than the third planet, Earth. The ancients thought Venus was two objects, the morning and evening "stars," until realizing that both never appeared in the same night. The Greeks named the planet Venus for the goddess of love and beauty.

In the skies of Earth, Venus is certainly beautiful—the brightest object other than Earth's moon. In 1610, Galileo found that, like the Moon, Venus has phases. This helped prove the scorned heliocentric, or sun-centered, theory of our planetary system, first proposed by the Greek Heracleides in 350 B.C. The particular phases of Venus meant it was shining by reflected light from the Sun, and doing it in such a way that Venus had to be circling the Sun, not Earth.

Venus is considerably less beautiful close up. Without water and hotter than Dante's hell, Venus is probably lifeless. Surface heat is said to be 730° K, or 850° F, hot enough to melt lead.

The atmosphere of Venus is 95 times as dense as Earth's. It's mainly carbon dioxide, with even more poisonous additions of sulfuric, hydrofluoric and hydrochloric acids. Early in 1992, R. David Baker II and Gerald Schubert, of UCLA, showed that in certain areas the atmosphere of Venus forms convection cells that are thin but horizontally large, possibly due to the dynamics of the atmospheric layers, now under investigation.

One reason scientists are trying to understand the Venusian atmosphere is that someday an attempt may be made to improve it (like seeding the clouds with bacteria and blue-green algae). Another reason is the threat of a "greenhouse effect" warming up Earth. Venus already has a runaway greenhouse effect, and if we understand that, we might be able to prevent Earth from a similar fate.

Using computer simulations, Matthew Newman and Conway Leovy, at the University of Washington, recently investigated the strong winds that rotate the Venusian atmosphere every four days, while the planet below takes 243 days to complete a rotation. Solar radiation on the dense atmosphere apparently drives the thermal tides, which promote the winds. Scientists are still trying to answer other questions about the Venusian atmosphere, such as: Why are the clouds over the poles so warm? What maintains the background rotation rate of the lower atmosphere?

The surface of Venus is under study, too. Pocked with craters, it features mountains (the largest is named Maxwell), rift valleys, volcanic domes and rivers of lava longer than any water river on Earth. It was thought that Magellan's images revealed a recent landslide; then this was said to be an error. More recent study indicates that there probably were Venusian landslides, but in the past.

Scientists thought Venus had no tectonic plates like those moving around on Earth's surface. In May 1992, Dan P. McKenzie of Cambridge University told the American Geophysical Union that Venus may have a modest, patchy variety of plate tectonics. The most compelling evidence is a big, smudgy shape named Artemis, in the equatorial highland called Aphrodite Terra. Artemis may be a region where new crust is formed. Next to Artemis are trenches similar to the subduction trenches seen on Earth's ocean floor, where plates descend under other plates.

Gerald Schubert, at UCLA, and David T. Sandwell, at the Scripps Institution of Oceanography, have also analyzed the Magellan images. They suggest that subduction on Venus is different. Earth's crust wells up from a midocean ridge, then moves long distances horizontally until it dives under another plate at the subduction zone. On Venus, plates don't seem to move horizontally. It's postulated that when the Venusian surface is torn by mantle heat, crust on either side of the tear sinks vertically as molten mantle rock rises.

Dark margins encircle half of the Venusian craters, but also exist without a central crater. Geophysicist K. J. Zahnle believes that many meteorites would break up as they plunge into the "aerodynamic stresses" of Venus's atmosphere, causing shock impacts similar to the blast which felled trees in Tunguska, Siberia, on June 30, 1908.

Unlike Mars, Venus and Earth are both planets with good-sized atmospheres. On Earth, water wipes away the evidence of much of what happens to our planet's surface. On Venus, we can detect what the surface is and was like, and how the strange Venusian atmosphere affects it.

A Martian Asteroid?

In 1990, a small new asteroid was spotted by Henry E. Holt and David Levy at the Palomar Observatory in California. The interesting thing about it is that it may be moving in the orbit of the planet Mars.

The story of such asteroids goes back to 1772, when a French astronomer, Joseph-Louis Lagrange, showed that there were five places where a small body could keep time with a planet as the planet circled the Sun.

These places are called "Lagrangian points," and they are

numbered L1, L2, L3, L4, and L5. Of these points, L1, L2, and L3 are unstable. Any object at those points, should it happen to move away from the point even slightly, would continue to move away and never come back. L4 and L5 are stable, however. If an asteroid is at one of those two points, then, even if it wanders off, it comes back again, so that it vibrates about that point, so to speak, and can remain there indefinitely.

L4 and L5 are points along the orbit of the planet. L4 is sixty degrees ahead of the planet; and L5 is sixty degrees behind the planet. In either case, if you draw imaginary lines from the asteroid to the planet to the Sun and back to the asteroid, you have an equilateral triangle, a triangle with sides of equal length.

Lagrange was just working theoretically. No one knew of any actual cases of asteroids in the L4 or L5 position of any planet. But then, in 1906, a German astronomer, Maximilian Wolf, discovered an asteroid (the 588th to be discovered, by the way) that was circling the Sun in the orbit of Jupiter. Pretty soon, other such asteroids were discovered, some keeping time with Jupiter in the L5 position, sixty degrees behind, and some in the L4 position, sixty degrees ahead.

Wolf had named the first asteroid of this sort "Achilles," after the Greek hero in the Trojan war. Other asteroids found in the L4 and L5 positions received the names of other Greeks and Trojans who fought in the war. As a result, the L4 and L5 positions have come to be known as "Trojan positions," and the asteroids in them are "Trojan asteroids."

To this day, the Trojan asteroids connected with Jupiter are the only ones known. There might be asteroids in the Trojan positions of Saturn, Uranus, and Neptune, but those planets are so far away that asteroids accompanying them about the Sun would be too faint to be seen, unless they were unusually large.

As for the planets closer to the Sun than Jupiter is, they are small and poor in satellites.

Naturally, the Trojan positions we are most interested in

are Earth's. Is there an asteroid or two sharing our orbit about the Sun but remaining always sixty degrees behind or ahead? Even more exciting would be asteroids in the Trojan positions with respect to the Moon. They would share the Moon's orbit about the Earth, remaining always sixty degrees before or behind.

An asteroid in Earth's Trojan position would be fully 93 million miles from the Earth. (It would have to be at the same distance as the Sun, since the asteroid, Earth, and Sun would form an equilateral triangle.) An asteroid in the Moon's Trojan position would be only 237,000 miles away (the distance of the Moon), and we could visit it with no more trouble than it has taken to visit the Moon. Less, for the asteroid would have no interfering gravity of its own.

The trouble is that, try as we might, we have not spotted any asteroids in the Trojan positions of either the Earth or the Moon. Some years ago, it was announced that there might be some thin dust clouds in the Moon's Trojan positions, but that didn't turn out to be so.

There are people who recommend that large space stations capable of holding ten thousand people be built in the Moon's L4 or L5 position. That would result in artificial Trojan satellites, moving about the Earth in the Moon's orbit. There are even a group of enthusiasts who support such a notion and who call themselves "The L5 Society."

Now we get to the new asteroid discovered by Holt and Levy. Its position happens to be near Mars's L5 point, so that it may be a new Trojan asteroid, the first to be found in connection with any planet other than Jupiter. Of course, it might not be a Trojan asteroid even if it is in the L5 position. It may have a completely different orbit that happens to intersect (or nearly intersect) Mars's orbit at a point that, on this one occasion, happens to be near the L5 point.

Therefore, the asteroid has to be kept under observation for a while so that its orbit might be carefully calculated. Once that is done, it should prove to be either a Trojan or not a

Trojan. If it is not a Trojan it will never be heard of again. If it is a Trojan, it will be famous. (Of course, if Mars has one, I'm going to feel bad that Earth doesn't have one, too. That's planetary chauvinism.)

Mars for Humans

The Association of Space Explorers (composed of those who have been in space) recently had its eighth planetary congress, and the theme was "To Mars Together." The "together" is crucial, for investigating and making economic use of the planet Mars should be a global effort by all terrestrial nations, working together. It will seem expensive, but solving problems in space exploration can also solve problems for us here on Earth.

Space exploration spends money *on Earth*, creating useful new industries and markets, with new jobs. The 25 billion dollars spent on the Apollo missions of the 1960s and 1970s eventually gave back twenty times that in remarkable advances useful on Earth. Because of the space effort, there was a tremendous leap forward in many fields, especially metallurgy, electronics, ceramics, and computer technology.

Exploration and settlement of Mars is not a fantastical notion. It's practical. It could eventually provide another home for humanity in case anything happens to Earth. Furthermore, as our resources dwindle and pollution becomes an increasing menace, we need to learn about adequate recycling, efficient use of space (especially for growing food), shielding from lethal radiation (remember the growing hole in our ozone layer?), and living harmoniously together because you can no longer run away to easy-to-settle frontier lands if you don't like where you are. Living on difficult Mars will quickly teach us these things.

New products to make Martian living possible will appear in Earth's markets quicker than if we'd dillydallied about making them only for Earth. In the past, wartime necessity galvanized invention and industry toward vital new products, but we no longer want that kind of motivation.

We may build a base on the Moon first (it's only three days away), but for settlement, Mars is much more of a real planet. Mars isn't as big as Earth (approximately six thousand kilometers less in diameter), and its year is 687 of our days long. Its gravity is much stronger than the Moon's, but only 38 percent of Earth's.

Unfortunately, the Martian atmosphere is 95 percent carbon dioxide, but if the domed settlements we'd build happen to spring a leak, we could make repairs without worrying that the air coming in would kill us immediately, as that on Venus would.

Astronomers at the University of Hawaii and Tel Aviv University suggest that icy comets impacting on Venus, Earth, and Mars contributed to the formation of atmospheres. We can't find out more about this on Earth, and only with great difficulty on Venus, but it should be easy to investigate on Mars. Others have speculated on the possibility of adding water to Mars by towing a comet there.

Humans will need water wherever they go. Can they find enough on Mars? The planet now has an average temperature of $-60°C$, with a low atmospheric pressure. This makes it impossible for liquid water to exist on Mars. But there's hope that water will be found.

Martian surface features indicate that the planet might once have had a lot of water, snaking across the landscape in huge channels. University of Arizona astronomers Jeffrey Kargel and Robert Strom recently examined fourteen-year-old Viking photographs of Mars and concluded that Mars once had an ice age, with glaciers composed of a mixture of water and carbon dioxide. Many scientists believe that the polar ice caps still contain water ice as well as frozen carbon dioxide.

At the Department of Earth and Space Studies of UCLA, David A. Paige has examined old data provided by the Mars Viking probe. While the Martian soil samples from one small area showed no evidence of subsurface water, Viking made thermal maps of three regions that have always been brighter than the rest of Mars: Tharsis, Arabia, and Elysium. These areas seem to be covered by a layer of fine dust that is a good insulator. Paige postulates that near-surface ground ice can and probably does exist in these areas.

Mars may not have developed life (the question needs further investigation), but it is an active planet. Dr. Baerbel Lucchitta of the U.S. Geological Survey has reexamined Viking photographs and found evidence of a landslide in action. What else is going on? We need to get better acquainted with Mars and find ways of using the planet to help ourselves.

Saturn's Rings

Saturn's rings are the most beautiful objects in the solar system. While the other outer planets have rings, those that Jupiter, Uranus, and Neptune possess are thin, dark, and unimportant in appearance. Saturn's rings are large, bright, and glorious.

Why is that? Luke Dones of the Canadian Institute for

Theoretical Astrophysics believes we just happen to live in a time when Saturn's rings are beautiful and that they are, little by little, vanishing.

There are apparently two phenomena that are leading toward the disappearance of the rings. In the first place, the satellites of Saturn are forever tugging at the rings and stealing orbital energy from the myriads of particles the rings are made up of. As a result, the particles making up the rings are slowly spiraling inward toward Saturn and, eventually, will disappear. Dones estimates that the disappearance will take about 100 million years.

A second process is the constant collision of the ring particles with comet dust grains. The comet dust breaks up the ring particles, making them constantly smaller, and their energy then more rapidly disappears. There is an estimate that the effect of the comet dust again would bring about the disappearance of the rings in 100 million years.

Furthermore, the comet dust is absolutely black in color so that if even a little of it is added to the ring material, the rings of Saturn will grow much dimmer. Yet there are parts of the rings that are bright and that are clearly made up of ice. This would indicate that the rings have not yet been exposed to comet dust for very long.

The same processes might affect the thin rings of Jupiter, Uranus, and Neptune, but in those cases, the particles might be replaced by material blasted off the satellites of those planets. Saturn's rings, on the other hand, cannot be replaced since they are too large.

The question is, where did the rings of Saturn come from? Dones suspects that a large comet, or several of them, must have skimmed by Saturn close enough to be ripped apart. Comets are made up of icy materials, so that the rings formed would be made up of ice.

Of course, questions arise. Why should comets be torn apart by Saturn, and not by the other outer planets, especially the much larger Jupiter? Furthermore, it can't be a single comet that would break up. Dones estimates there would have to be between ten and one hundred comets. Why Saturn and not the others?

Someday, a Saturn probe will be in a position to study Saturn's satellites and see if they are slowly moving outward because they are stealing the ring's energy. In other words, they move outward as the rings move inward toward Saturn. That would tend to settle matters.

Saturn has also a large number of satellites, and some of them have turned out to be unusual. Charles Yoder of the Jet Propulsion Laboratory studied the innermost satellites, Janus and Epimetheus. They lie just beyond the edge of Saturn's rings and were not discovered till 1966, when Saturn's rings were seen edge on from Earth.

As it turns out, Janus and Epimetheus, which are small satellites, have almost identical orbits. Every four years they pass very close to each other and swap orbits. One ends up being a little closer to Saturn, the other a little farther. Yoder studied the manner in which the satellites switched orbits and decided that the satellites must have densities of less than 0.7 grams per cubic centimeter. This is considerably less than the density of the other satellites and is less than the density of pure ice.

Apparently, these satellites are merely piles of icy debris, with some 30 percent of their structure consisting of empty space. Is it possible that these satellites are actually conglomerations of ice particles from Saturn's rings? That is not so unbelievable. Janus is 220-by-160 kilometers in size, and Epimetheus is 140-by-100 kilometers. These and three other small satellites that just skirt Saturn's rings—Atlas, Prometheus, and Pandora—may all be coagulations of particles from Saturn's rings. They are small enough for the purpose, and may also be contributing to the disappearance, eventually, of Saturn's rings.

Of course, 100 million years is a long time on a human scale, and we need not be fearful that the glorious rings will disappear from our vision. Yet, even so, there is the sadness that sweeps over us when we think that something so marvelous is not forever.

Titan's Atmosphere

Atmospheres are an interesting phenomenon. Huge worlds with strong gravity can hold on to gas molecules and keep them from escaping into space. Thus, the Moon and Mercury, small worlds, do not have an atmosphere, and Mars has only a thin one. Earth and Venus have thick atmospheres.

Strong gravity is only one way in which an atmosphere can be held. The colder a world, the more slowly the molecules of gas about it can move and the more easily they can be held to the surface. The satellites of Jupiter, even though four of them are quite large, are still too warm to hold an atmosphere. Titan, the largest satellite of Saturn and the second largest in the solar system, is much colder and it can hold an atmosphere. The still colder worlds of Triton (Neptune's satellite) and Pluto also have atmospheres, but they are thin. Titan's is thick, denser than Earth's.

When the atmosphere of Titan was first detected by Gerard Kuiper in 1944, it seemed that the atmosphere was only 1 or 2 percent as dense as Earth's, and was made up of a thin layer of methane, a very common gas. The trouble is that methane is an easy compound to detect. If other gases were present that were hard to detect, the logical candidate would be nitrogen.

It was not till the time of the planetary probes, however,

that Titan could be observed closely. In a way it was disappointing, for it seemed only a featureless orange globe with nothing visible because the atmosphere was so hazy. Probes, however, entered Titan's atmosphere and sent back the astounding news that it *was* largely nitrogen, maybe up to 90 percent nitrogen. It was that which made Titan's atmosphere so dense.

This is interesting because only Titan and Earth have atmospheres that are largely nitrogen. The giant planets have atmospheres that are mostly hydrogen. Mars and Venus have atmospheres that are mostly carbon dioxide. Only Titan and Earth are unusual in this respect.

Where on Earth (or, rather, where on Titan) did the nitrogen come from? One possible answer arises from the internal structure of Titan. The central portions of Titan are rocky in nature, but around that rock is a thick, thick layer of ice—diamond-hard because of the low temperature.

It has been suggested that in Titan's earliest days, as the ice layer formed, it trapped nitrogen (which, apparently, is easily trapped under such conditions). Then, over billions of years, the nitrogen leaked out and formed the atmosphere. That doesn't really answer the question. Where did the nitrogen come from to be trapped?

Another suggestion: Saturn, like the other giant planets, has considerable ammonia in its atmosphere. Ammonia is made up of nitrogen and hydrogen. If Titan picked up ammonia from Saturn, then ultraviolet light would break it up into plain nitrogen and plain hydrogen. Hydrogen is made up of very small atoms. The smaller the atom, the faster it moves, and Titan couldn't hold on to it, but it could hold on to the heavier atoms of nitrogen.

The problem is that Titan would have to be considerably warmer than it is for such a reaction to take place. Well, perhaps Titan was warmer in very early times, but we don't know—so this remains a problem for astronomers to play with.

Recent studies by radar show that the reflections vary as

Titan turns, and, apparently, the best explanation is that Titan's surface is partly solid and partly liquid. The solid surfaces are continents of hard ice. What is the liquid composed of?

The methane in Titan's atmosphere is easily altered by ultraviolet light and converted to ethane, which is a kind of double molecule of methane. Methane remains a gas even at Titan's low temperature, but ethane is a liquid, so the current thinking is that Titan has a large ethane ocean.

This is very interesting because ethane is sort of midway between natural gas and gasoline. It burns very nicely and supplies energy just as oil does. In fact, we might well decide that Titan is the largest oil well in the solar system. Naturally, one imagines ourselves scooping up the ethane and carting it off for use elsewhere. It would be a supply that would last as long as the human race does.

But there's a catch. (There is always a catch.) Titan is so far away that to go out there, collect the ethane, and bring it anywhere in the inner solar system where it might be useful would result in absolutely prohibitive expense. Maybe the time will come when we figure out a way to do it economically.

Naming Neptune's Satellites

Voyager 2, in its close approach to Neptune in 1989, spotted six small satellites close to the planet, and four of these have now had names supplied for them, to be made official at the International Astronomical Union's meeting in Buenos Aires.

Astronomers weren't always intent on naming satellites.

For about three-fourths of a century, Jupiter's satellites, beyond the first four, were simply known in the order of their detection—Jupiter V, Jupiter VI, and so on, all the way to Jupiter XIV. It was only when rocket probes began to study planetary satellites in some detail that names began to be supplied for all of them.

In 1846, shortly after Neptune was discovered, a satellite was found to be circling it. Neptune, because of its greenish color, was named for the Roman god of the sea. The Greek equivalent of the name was Poseidon. The satellite was, appropriately enough, named Triton, after a son of Poseidon in Greek mythology. Triton was viewed as a being with the head and torso of a man and the tail of a dolphin.

Triton is a sizable satellite and for a hundred years was thought to be larger than our Moon, because it was thought to have a dull surface and therefore had to be large to reflect as much light as it did. Voyager 2, however, found it had a shiny surface, so that it reflected the necessary light even though it was distinctly smaller than our Moon.

For a century, Triton was the only satellite Neptune was known to possess. In 1949, however, a far smaller satellite was found with an eccentric orbit far out from Neptune. It revolved about Neptune, by sheer coincidence, in 365 days, the same time it took Earth to revolve about the Sun. It was named Nereid, which is not really the name of a mythological personage, but of a group of them. The Nereids were sea nymphs, the fifty daughters of a sea god named Nereus.

By the 1970s and 1980s, it was taken for granted by astronomers that Neptune had additional satellites circling the planet at close distances. The probes had discovered such satellites circling Jupiter, Saturn, and Uranus. They were not visible from Earth, firstly because they were small and therefore very dim, and even more important, because they were so close to the planets they circled that the light of those planets washed them out.

And sure enough, when Voyager 2 passed Neptune, six close-in satellites were spotted.

One of them was a bit larger than Nereid, and so it became the second largest of Neptune's satellites, while Nereid dropped to third place. That Nereid was spotted from Earth while the new larger satellite was not was entirely because Nereid was far enough from Neptune to be seen.

The new large satellite, which is, on the average, about 250 miles across and is about 65,000 miles from Neptune's center, is being named Proteus. Proteus is an interesting mythological character. He is a herdsman for Poseidon, watching over the sea god's flocks, which are seals. He is supposed to have the appearance of an old man who can foretell the future and will do so for you, if you sneak up on him while he is sleeping and hold him fast. This is not that easy because Proteus can change his form into that of a wild animal, into fire, and into other things you must have considerable fortitude to continue holding. If you have that fortitude, Proteus eventually gives up, returns to his proper shape, and tells you your future.

Proteus was the first of the new satellites to be discovered. The third is about 90 miles across and is about 32,000 miles from Neptune's center. The name suggested for it is "Despina," which is not very appropriate, for it has nothing to do with the sea. The word means "the mistress," and the Greeks used it for Aphrodite, the goddess of love, for Demeter, the goddess of agriculture, and for Persephone, the goddess of the underworld. No hint of the sea.

The fifth of Neptune's newly discovered satellites is about 50 miles across and is about 30,000 miles from Neptune's center. It is named "Thalassa," which, unlike Despina, is a very good name. It is the Greek word for "ocean," and what better name for a satellite of the sea god?

Then there is the sixth of the satellites, which is about 40 miles across and is about 28,000 miles from Neptune's center. The name suggested for it is "Naiad." Naiad, like Nereid, is the name given to a group of mythological beings. The naiads are water nymphs, but they were supposed to rule over the fresh waters of the Earth—rivers, brooks, springs, and fountains. They were usually pictured as young and beautiful women leaning on urns from which water pours out.

The second and fourth of the newly discovered satellites of Neptune have not, as yet, been assigned names. I don't know why, but you can be sure they will not be missed. Why not Scylla and Charybdis? These are two sea monsters, the first a kind of octopus, the second a kind of whirlpool, encountered by Odysseus in the *Odyssey*.

Triton, the Last Satellite

There are seven large satellites that we know of in the solar system, and now astronomers have seen them all close up. Our own Moon has been studied telescopically for nearly four centuries. The four large satellites of Jupiter—Callisto, Ganymede, Europa, and Io—together with the large satellite of Saturn—Titan—have, in the course of the last decade or so, been studied close up by probes.

Until 1989, however, Neptune's satellite Triton the last (and farthest) of the seven, had still been seen only as a point of light through the telescope. That magnificent probe, Voyager 2, passed within 24,000 miles of Triton and was able to photograph it at close range.

It had been suspected that Triton would resemble Saturn's satellite Titan in appearance, but that proved to be quite wrong. The big difference was this: Titan has a thick atmosphere of methane and nitrogen. The methane is acted on by light from the distant Sun and forms larger hydrocarbon molecules that spread as a fog of liquid droplets throughout the atmosphere. The cameras of Voyager 2 could not penetrate the fog, and Titan's solid surface was never seen.

Triton, on the other hand, is three times as far from the Sun as Titan is, and is therefore considerably colder. It is in fact the coldest object astronomers have yet studied. Triton also has an atmosphere of methane and nitrogen, but most of it is frozen out, leaving behind only a thin atmosphere of vapors through which the surface can be easily seen. That surface, it turns out, is slick with frozen methane and nitrogen, especially in the southern hemisphere.

That slickness is important. Until now, the only way of judging the diameter of Triton was by measuring the amount of light it reflected. It was assumed that the intensity of reflection was equal to that of the other large satellites. From that, astronomers could calculate the size Triton would have to be to reflect enough light to appear as bright as it does when seen from Earth. The best guess was that its diameter was about 2,175 miles, making it just a tiny bit larger than our Moon.

However, with its smooth surface covered by shiny, frozen gases, Triton reflects light much more intensely than had been supposed. Under those conditions, it would have to be smaller to reflect the amount of light we see—and smaller it is. It turned out that Triton is only 1,700 miles across, making it the smallest of the seven satellites.

Still, it is the most colorful. The surface has pink regions where the Sun has forced the methane into larger and more complicated molecules. It is also bluish, where sunlight is reflected from tiny crystals, with the same scattering effect that gives our sky its own beautiful blue.

The most interesting thing about Triton, however, is the odd variations in its surface structure. It has ridges, grooves, and all sorts of irregular shapes, but very few of the craters that mark most of the other smaller bodies of the solar system.

It must have developed craters in the first billion years of its lifetime when it was bombarded by various sizable bodies that were coming together to form the planets and satellites of today. After that, though, Triton must have melted and then refrozen smoothly.

What melted it? We don't know. Perhaps it collided with another of Neptune's satellites. Perhaps the shock melted both bodies and combined them. Perhaps that is why Triton revolves about Neptune in the wrong direction. All the other large satellites revolve about the planet in the same direction the planet turns on its axis, but Triton revolves in the direction opposite to that of Neptune's rotation.

If nothing had happened to Triton after it had refrozen, its surface would be completely smooth, but there *are* various irregularities. That makes it different from Jupiter's satellite Europa, which is completely smooth because it is covered entirely by an ice glacier that refreezes whenever it is struck, melted, and broken by a meteorite, so it remains smooth. Triton is more like Jupiter's satellite Io, which is volcanic. The melted rock that pours out of Io's volcanoes fill up the craters and leave the surface smooth, except where the volcanoes are actually active.

Triton also seems to be volcanic, but in Triton's case there is no rock to speak of in its outer layers, and no heat source powerful enough to melt it if there were. Instead, Triton's temperature, while extremely low, is still warm enough (especially in places where heat reaches the surface from the warmer interior of the satellite) to melt and vaporize the nitrogen.

The nitrogen blasts outward, melting some of the water-ice that surrounds it. The ice refreezes quickly, forming ridges and mounds. These ridges are up to a few hundred yards high and can sometimes run for hundreds of miles along the surface. All the variety of Triton's surface may be the result of "ice volcanoes," the only ones of this kind that we know of in our solar system.

The Solar System's Greatest Storm

In 1990, two amateur astronomers, Stuart Wilbur and Alberto Montalvo, independently noted the first signs of what turned out to be the greatest atmospheric storm ever seen anywhere in the solar system. And it wasn't on Jupiter.

This, in itself, is a source of surprise, for Jupiter is by far the largest of the four "gas giants" of the outer solar system. In addition, it is the closest of the four to the Sun and gets the most energy from it. Not only that, but it rotates more rapidly than any of the others, which puts the atmosphere into violent motion. This combination of a great influx of energy, a tearing spin, and a mighty gravitational pull makes Jupiter apparently the most active of the planets. Its atmosphere is torn by enormous storms that whip across it from west to east, and these storms show themselves as varicolored belts interrupted by cyclonic spirals. The largest of Jupiter's storms is the "Great Red Spot," a kind of hurricane that has been continuing for centuries and that spreads over an area into which the entire Earth could be comfortably fitted.

Saturn is farther from the Sun than Jupiter is; it is distinctly smaller; and it rotates a little more slowly. All this would lead us to suspect that Saturn's atmosphere would be quieter and less turbulent than Jupiter's—and it is.

Uranus is still farther from the Sun, still smaller, still slower in rotation, and so it should be still quieter—and it is. In fact, Uranus is a placid planet indeed with scarcely any atmospheric features we could understand.

In 1989, when Voyager 2 observed Neptune close up—

that planet being a virtual twin of Uranus but even farther from the Sun—astronomers expected to see a planet as quiet as Uranus. Instead, they found Neptune's atmosphere to be raging in a fashion very much like that of Jupiter's. It even had a "Great Dark Spot" very much like Jupiter's Great Red Spot. Where does Neptune get the energy for this? It's a question that is fascinating astronomers.

But what Wilbur and Montalvo noted on September 24 wasn't on Neptune, either. It was a small white spot on Saturn. This was not in itself unusual. Saturn revolves about the Sun in about twenty-nine and a half years, and during that revolution, it passes the point at which its north pole is tilted as closely to the Sun's direction as possible. This is the point that is equivalent to the "summer solstice" on Earth. At this time, Saturn's northern hemisphere gets its most intense supply of energy from the Sun, and that means that storms are more likely. As a result, every thirty years or so, white spots are observed on Saturn.

In the fall of 1990, Saturn was passing through its summer solstice, and so storms were expected. The initial reports of the white spot might have seemed routine, therefore.

But then came the surprise. The white spot began to grow in a totally unprecedented manner. Three days after its discovery, it had become a very bright oval. After a week had passed, it had spread out over a width of 10,000 miles, and after a month had passed, it was 50,000 miles wide. By October 23, the "Great White Spot" completely encircled Saturn and was the greatest atmospheric storm ever observed in the solar system. Nothing like it had ever been seen before.

Perhaps the most satisfactory part of the discovery was when the Hubble Space Telescope (HST) took photographs of Saturn that showed the storm in far greater detail than anything Earth-based could do. The great detail in HST's photographs enabled astronomers to study the storm carefully and to see how it developed and slowly changed. Other white spots appeared and dark lines, too. The details of the storm match what scientists have felt ought to happen, and that is highly satisfactory, too.

What does the storm consist of? The general feeling is that it is fed by upwellings of gas from below. Ammonia, in which Saturn's atmosphere is rich, moves upward and freezes in white crystals, so what we are seeing is thousands of miles of frozen ammonia.

Calculating
a Satellite

Recently, Mark Showalter of the NASA-Ames Research Center in California discovered a new satellite circling Saturn and did it in an unprecedented way. He studied close-up photographs of the rings taken in 1980 and 1981, noticed certain wavelike patterns, deduced from that that an unknown satellite must be present, calculated where it ought to be, looked—and there it was.

The discovery of satellites of the planets (not counting our own Moon, which has been known since human beings started looking at the sky) began in 1610. In that year, the Italian scientist Galileo turned his newly built telescope on Jupiter and found four sizable satellites circling it. These are Io, Europa, Ganymede, and Callisto.

In 1654, an equally large satellite, Titan, was discovered circling the planet Saturn. Before the century was over, four more satellites, of lesser (but still considerable) size, were discovered to be circling Saturn.

Such discoveries continued. As telescopes grew better and better, smaller satellites that were farther off could be seen. Soon after the planet Uranus was discovered in 1781, William Herschel, its discoverer, spotted two satellites of intermediate size circling it. A half century later, two more were discovered. Again, soon after Neptune was discovered in 1846, a sizable satellite, Triton, was discovered to be circling it.

No additional large satellites have been discovered since 1846, and very likely, no more exist. However, a drizzle of discoveries continued to be made of smaller satellites.

The most dramatic of such discoveries came in 1877, when Asaph Hall examined the near neighborhood of Mars in the search for small satellites. Finally, he gave up, but Mrs. Hall said, "Try it one more night, Asaph." He did and discovered Mars's two satellites, Phobos and Deimos.

In 1892, E. E. Barnard discovered a fifth satellite of Jupiter, much smaller than the first four and one that circled Jupiter at a much shorter distance. It was the last satellite to be discovered by eye. Since then, all new satellites have been discovered by means of photography.

These discoveries continued right on into the twentieth century. Even as late as 1948, G. P. Kuiper discovered a fifth satellite of Uranus, and in 1949, he discovered a second satellite of Neptune. Both were quite small. In 1978, there was a true surprise when J. W. Christy discovered that the tiny planet Pluto had a satellite almost as large as itself.

As the age of rocketry dawned, then, here was the score of known satellites in the solar system. Mercury had none. Venus had none. Earth had one (the Moon). Mars had two tiny satellites. Jupiter had twelve (four large and the rest quite small). Saturn had nine (one large and several middle-sized). Uranus had five (four of them middle-sized). Neptune had two (one large). Pluto had one. That made thirty-two satellites altogether.

Then, however, beginning in 1979, the Voyager probes zoomed past the outer planets, and their cameras detected tiny satellites that couldn't possibly be seen from Earth. Three were discovered to be circling Jupiter; no fewer than nine were found to be circling Saturn. Ten were found to be circling Uranus, and several new ones were found to be circling Neptune. The total number of known satellites is now approaching the sixty mark.

The most magnificent set of satellites belongs to Saturn. Not only were there seventeen altogether, but several of them share orbits. The middle-sized satellite Tethys has two tiny satellites sharing its orbit, one preceding it and one following it. Dione has one tiny satellite sharing its orbit. Saturn also has a magnificent ring system, broad and bright. Jupiter, Uranus, and Neptune also have rings, but they are thin, dark, scanty, and almost contemptible. Why Saturn should be so spectacularly endowed we don't know.

Saturn's rings are composed of tiny bits of ice that are distributed almost evenly over their broad width. There are, however, gaps in the rings. The largest is known as Cassini's division, and the next largest is Encke's division—both named for the astronomers who first noted them. The gaps are the result of the gravitational influence of those Saturnian satellites that are nearest the rings, so that at certain distances the bits of ice are pulled outward or pushed inward.

Voyager photographs showed that Encke's division had wavy edges. Showalter thought this must be the influence of a satellite actually inside the rings. He used a computer to determine where the satellite must be to produce the waves. He then studied those portions of the rings, magnifying them by modern techniques—and there was the eighteenth satellite of Saturn. It was only twelve miles across, but it was much larger than the icy fragments that made up the ring. It was the first satellite to be found by computer.

The Strange Comet

There's a strange object that circles the Sun between the orbits of Saturn and Uranus, an object that interests astronomers more and more these days.

It was discovered in 1977 by Charles Kowal of Cal Tech. He named it Chiron, and it was thought to be an asteroid. It is about 150 miles in diameter, which makes it rather large for an asteroid, and it circles the Sun far beyond the ordinary asteroid belt.

That would be enough to make it strange, but it is at the near end of its orbit now, and astronomers, studying it closely with modern instruments, determined in 1988 that it was surrounded by a layer of dust and gas, which meant it was not an asteroid, but a comet. That makes it stranger than ever, for it is the hugest comet ever seen, with about 10,000 times the mass of the famous Halley's comet.

Comets in general have become more controversial lately. There are two kinds of comets. There are the "long-period comets," which circle the Sun in periods of thousands of years; and "short-period comets," which circle it in less than 200 years. Halley's comet and Chiron are both short-period comets.

The long-period comets come in toward the Sun from the far, far reaches of space, over a thousand times as far away as the farthest planet. What's more, they come in from all possible directions. As a result, astronomers are pretty sure there is a large sphere of many billions of cometary bodies surrounding the Sun. This is called the "Oort cloud," after the astronomer who first suggested its existence. Every once in a while one of these bodies, affected by the gravity of distant stars, drops down into the inner solar system.

For a long time, it was thought that every once in a while one of these long-period comets passed close enough to a planet, particularly giant Jupiter, to have its orbit changed by gravitational attraction. That comet might then be "captured" and remain permanently in the planetary system, and would thus become a short-period comet.

The short-period comets don't come from every direction, however, but circle the Sun pretty much in the plane that the planets use. This did not ruin the theory, for it was assumed that in capturing a long-period comet, the planets forced it into their own orbital plane.

What does mess up that idea, now, is that recent computer simulations of the situation show that long-period comets are very hard to capture. One or two might be trapped, but there are 150 short-period comets, and that many captures seems far out of the question.

There is now, therefore, a suggestion that there is a second cometary region around the Sun, one that is much closer than the Oort cloud and, indeed, exists not as a sphere but as a narrow belt. It is called the "Kuiper belt," after another astronomer, and it may be the source of the short-period comets.

The astronomer Mark Bailey, of the University of Manchester, has suggested something rather startling. It is that very few comets enter the solar system from the Kuiper belt, maybe only one or two, but they are giant comets like Chiron.

By means of a computer, he followed the orbit of Chiron for 100,000 years into the past and found that it is an unstable orbit moving much nearer the Sun at some times than at others. He suggests therefore that once in the far past, when Chiron passed quite near the Sun, it fragmented and gave rise to all the short-period comets.

This is not impossible. All the short-period comets, other than Chiron, if taken together would have only about 2 percent of Chiron's mass. That means that after all the fragmentation, 98 percent of Chiron is still there.

This is an interesting idea, indeed, but I find it hard to take. Some short-period comets, like Biela's comet, exist entirely as lumps of icy substances mixed with dust, and after approaching the Sun a number of times they can dissipate altogether into a dust cloud, as, indeed, Biela's comet did. Other short-period comets, like Encke's comet, have rocky cores, so that by now almost all the ice and dust are gone and only the rocky core remains, leaving Encke's comet almost like an asteroid.

It's difficult to see how Chiron can fragment and produce some comets with rocky cores and some without.

Another thing: Does the Kuiper belt really exist? The Oort cloud is so far out that it is hopeless to try to detect it, but the Kuiper belt should be detectable. Several probes have passed beyond the orbit of Neptune and have detected no cometary bodies. Still, even if the belt exists, the comets would be very far apart and might easily go undetected by a probe passing through it at some particular spot.

What we need is a probe that goes into orbit about the Sun just beyond Neptune. It should circle the Sun in the direction opposite to that in which the planets (and presumably the cometary bodies of the Kuiper belt) do. It might, then, every once in a while in its two-hundred-year journey about the Sun, spot the approach of a giant comet like Chiron.

More
on Comets

Worries about the present, near future, and the next century are bad enough, but recent headlines proclaim peril from the skies in August 2126. The culprit will be a comet.

So, let's look at comets. They've been titillating the human mind since prehistory, and they are intrinsically interesting and beautiful. They're useful—studying them helps us understand the solar system, and someday we might use their ice to supply water to colonies on Mars.

Comets are objects usually not much larger than a kilometer across, surrounded by a "coma" of dust and gas. The cometary body is composed of dirty ice or frozen mud around a rocky core.

The frozen mud theory is recent, based on infrared studies which indicate that the ice of comets is much more mixed with "dirt" than had been thought. This links comets with similar outer bodies of the solar system like Pluto and Triton, which some scientists now believe may have been formed from crunching together many comets.

The solar system's cometary shell is inhabited by many billions of comets, all originating five billion years ago at the same time when the rest of the solar system was condensing out into sun, planets, moons, and asteroids.

Of course, there are always exceptions. Comet Yanaka has recently been analyzed and doesn't seem to have carbonaceous matter. If it was formed along with our other comets, it upsets theories about the composition of the nebula that gave rise to our solar system. Instead, scientists postulate that it originated within a cloud of interstellar molecular matter, and was later captured by our solar system.

As comets approach the Sun, they develop "tails"—made of dust, vaporized ice, and a "plasma" of electrons and molecular ions. Sometimes the components of the tails are present separately, in which case the dust tails curve and the "plasma" tails are straight. Under pressure from the solar wind, the tails always point away from the Sun, a fact noted four and a half centuries ago by Girolamo Fracastoro and Peter Apian.

So many comets, many with magnificent tails, are seen that it's obvious comets don't all stay in the far-off orbit at the edge of the solar system. The orbits of many comets are distorted by collisions with other comets and asteroids, or the pull of gravity from the giant planets, even other stars.

Some comets escape from the solar system completely; others swing closer to the Sun, where they lose mass. Halley's was found to lose 30 tons per second. Although the Giotto spacecraft reported that Comet Grigg-Skjellerup is losing only 100 kilograms per second, that's still more than anticipated.

Of the comets that come into our stretch of the solar system, the smallest orbit belongs to Comet Encke, making its trip around the Sun in 3.3 years. It gets almost as close to the Sun as Mercury does. Other comets come into the inner solar system only once in many years—76 for Halley's; 21,7000 for Comet Kohoutek; millions of years for others.

NASA has just begun the International Ulysses Comet Watch, while the Ulysses spacecraft mission studies the Sun (in 1994 Ulysses will cross the Sun's south pole). It's hoped that during this time amateur astronomers will photograph comets as they appear. The data from comet photographs and from Ulysses will aid study of the solar wind, whose charged particles affect Earth, too.

Sometimes it's hard to tell a "dead" comet from an asteroid. Recently astronomers at the European Southern Observatory decided that Asteroid 4015 is probably the same as Comet

Wilson-Harrington, which seems to have lost its ice cover since its discovery in 1949. Perhaps there are lots of dead comets around.

Impacting comets are thought to have contributed material to Earth. Since water, ammonia, and hydrogen cyanide (present in most comets) form adenine, one of the components of DNA, it's been postulated that the organic chemistry that led to life on Earth was either started or speeded up by comets.

There's even a current theory, much disputed, that there are still lots of midget comets nearby, leaving mysterious spots on ultraviolet images of Earth's atmosphere, and depositing water on Earth when they come in.

And now, about the big one, the comet that may be a major headache for our descendants, according to the warning just issued by the conservative International Astronomical Union. It is Comet Swift-Tuttle, first seen in 1862. It was seen again on September 27, 1992, by Japanese astronomer Tsuruhiko Kiuchi, and seems to have improved the Perseid meteor display (caused by the debris previously left by the comet).

Comet Swift-Tuttle's orbit will bring it closer next time, and the next—and chances are 1 in 10,000 that it will collide disastrously with Earth in 2126. By that time the human race will have spaceships patrolling the solar system, ready to shoot down or shove out of the way dangerous objects.

If our technology can't handle the comet by then, it means we've destroyed our civilization, possibly our existence, before the comet has a chance to do it for us. So don't worry about the comet, just uphold civilization!

Our Own
Private Sun

The sun, whose rays are all ablaze with ever-living glory,
Does not deny his majesty—he scorns to tell a story!
　　He don't exclaim, "I blush for shame, so kindly be indulgent."
But, fierce and bold, in fiery gold,
He glories all effulgent!

When *The Mikado* was first performed in 1885, most people, including lyricist W. S. Gilbert, accepted the heliocentric version of the solar system. Although it had been 342 years since Copernicus's book was banned by the Catholic church, a mere 50 years had passed since the ban was lifted. Of course, the Greek philosopher Aristarchus had decided in 280 B.C. not only that the Sun was bigger than the Earth, but that all the planets revolved around the Sun. (He couldn't prove it, and nobody paid any attention.)

In *The Mikado*, Yum-Yum believed that the Sun's glory was ever-living, but we know now that our Sun is about 4.7 billion years old, halfway through its lifetime. Nevertheless, another 4.7 billion years gives us plenty of time for investigating the Sun and deciding what to do when it expands to a red giant, engulfing Earth. Since we can't wait around until the red giant Sun collapses into a white dwarf, the most sensible thing to do would be to leave, in enclosed, powered space colonies to which we've already learned to adapt.

Although the "majestic" Sun is only a rather solitary "main sequence yellow dwarf" star out on a galactic spiral arm, it is Earth's own private star, whose radiation affects no other living planet. At 150 million kilometers from us, our Sun produces

149

the heat and light that make life on Earth possible. Plants use the Sun's energy through photosynthesis. Animals use it by eating plants or each other.

The dangerous practice of burning fossil fuels could be replaced by the use of solar energy. New technologies have markedly improved on the photovoltaic cell, which was invented in 1876 to convert sunlight to electricity. Solar cells and panels are now more efficient and getting better all the time, using thin films or silicon panels of various types. I use a pocket calculator powered by light, so I never have to change a battery.

The Sun is truly "fierce and bold." Inside it, hydrogen is converted to helium by nuclear fusion with a temperature that may be as high as 15 million°C at the core. The seething, golden, visible surface of the Sun, called the "photosphere," is covered by the thin chromosphere, from which most ultraviolet radiation comes. No one is sure why the chromosphere is much hotter than the surface below it.

Next is the eerily beautiful corona, even hotter and visible to us only during solar eclipses. The total "atmosphere" of the Sun extends out to the farthest reaches of the solar system and is under intensive study, but only lately has the "body" of the Sun itself revealed some of its secrets.

Dr. David H. Hathaway, at the National Solar Observatory, recently found that the gases of the Sun's "body" have a definite direction of flow. They move along the outer layer of the Sun from its equator to the poles, then back again at a deeper layer, from which gas surges up again to join the surface flow. Hathaway thinks the flow carries magnetic fields that produce sunspots and solar flares.

Flares emerge from the Sun's corona, to be seen in every eclipse photo. A Japanese probe launched in August 1991 confirmed the causal connection between sunspots, solar flares, and increased production of solar X rays. The probe showed that flares last longer than scientists previously believed, probably by a self-sustaining process of looping patterns of flow in magnetic fields. A flare may cause a nearby loop to release its energy so that groups of flares expand in violence. According to orbiting observatories, an increase in gamma rays is associated with a flare, whose magnetic field stimulates nuclear particles.

Charles Lindsey and several other solar astronomers studied the Sun during the 1991 total eclipse. Using the "submillimetre-wavelength" James Clerk Maxwell Telescope on Hawaii's Mauna Kea, they were able to measure the height and temperature of the chromosphere more accurately. They found that "magnetically entrained funnels of gas" do indeed get hotter above the Sun's surface, but not as hot as had been previously believed. Lindsey et al. have also monitored surface vibrations. They found "shadows" on the Sun, streaks of wave motion that seem to connect magnetically active areas like sunspots and may be emerging from deep down.

It's important to discover more about sunspots and their cycles (every eleven years). When sunspots are most active, the resulting increase of radiation can cause havoc to electronic devices on Earth (and in orbit). The danger to us is sufficient to warrant more investigation of our Sun and all its glory.

Out of
the Sun

Our lives are inevitably affected by what solar astronomers call the "Sun-Earth interface," but most humans are not aware of anything coming out of the Sun except light. The ultraviolet side of the spectrum of light is invisible to us, even as it gives us sunburns and skin cancer. We also can't see infrared, the 60 percent of light at the other end of the spectrum. The Sun, however, emits not only light but X rays, charged particles, and neutrinos.

Charged particles zoom out of the Sun in what's called the "solar wind," reaching Earth in three and a half days. During intense solar flare activity, the solar wind can increase to the dangerous levels we call a "solar storm."

When the solar wind strength is up, lower latitudes of Earth witness the aurora borealis, the beautiful northern lights. Since 1989, when magnetic storms displaced important satellites and knocked out power to Quebec, scientists have tried very hard to improve prediction of such storms. The best bet right now is to monitor the solar corona, noticing when masses of it tear away into space and increase the solar wind.

Because our ultimate future may depend on space exploration and settlement, it's important to learn as much as possible about the solar wind. STEP (Solar-Terrestrial Energy Program) is a seven-year project coordinating research results from observatories, satellites, and computer simulations. Since the Sun changes from moment to moment, so does what comes out of it, changing the transfer of solar energy in the interface with Earth. All these variations in the solar wind must be observed and measured.

Studying the solar wind also helps scientists understand other stars. The universe itself may become easier to understand, for ionized gases like the solar wind apparently constitute most of the known matter in the universe.

One particle coming out of the Sun is the neutral neutrino, a tiny object but a perennially big puzzle. In the past, scientists believed that the flood of neutrinos from the Sun was an exact measure of the nuclear fusion presumably going on inside the Sun.

Twenty-five years ago, when solar neutrinos were first counted, scientists were startled to find fewer than had been expected from the "standard model" theories about how main sequence stars like the Sun work. Scientists worried that the standard model would have to be discarded. There seemed to be three possibilities.

One: Neutrinos may not actually be produced in big quantities by the Sun's proton-proton fusion reaction.

Two: Neutrinos may be produced in the expected big quantities, but only a few can be counted because on their way to Earth something happens to them.

Three (hotly debated): The energy produced by nuclear fusion in the Sun's core may somehow be more efficient than was thought. This would mean a lower temperature at the core, which permits lower numbers of neutrinos.

Proton-proton fusion reactions in the Sun's core produce the electron neutrino, which comes in two speeds—high and low energy. Until recently, "neutrino detectors" could register only high-energy neutrinos. Raymond Davis, in the 1960s, used tanks of carbon tetrachloride placed deep in mines. Newer projects featured tanks of gallium. One of the gallium tanks has recently been increased in size. This tank has found some low-energy electron neutrinos, but still not enough. Physicist S. Bandler at Brown University proposes a new detector made of silicon wafers plus superfluid liquid helium to find more of the low-energy neutrinos.

Another theory of neutrinos is attracting attention. The "M.S.W." theory is named for the Russian and American scientists Stanislav P. Mikheyev, Alexi Smirnov, and Lincoln Wolfenstein. According to their theory, solar nuclear reactions produce enough neutrinos, but they oscillate between three forms of neutrino: the known and measured electron neutrino, and the muon and tauon, found only in laboratory experiments.

If the M.S.W. theory is correct and electron neutrinos do change form, there may actually be enough solar neutrinos to confirm standard theories about solar fusion, but we don't see the expected form because somehow the electron neutrinos have changed form on the way to Earth. The problem is to build a detector that will find the solar muon and tauon neutrinos.

If scientists do prove that there is oscillation between the three forms of neutrinos, such change of neutrino form would indicate that neutrinos possess some mass. Cosmology will be galvanized! There are so *many* neutrinos, coming out of the huge number of stars in all the galaxies of our universe.

If those neutrinos aren't totally weightless, we may have the answer to a fundamental question about the universe— how will it end? Umpteen megabillions of slightly "weighty" neutrinos would provide enough "critical mass" in the universe to ensure its eventual collapse.

Then the whole Big Bang scenario can start all over again!

Cosmic Danger

Cosmic rays are produced by violent events in deep space, such as the explosion of stars or the activity of black holes. They consist of extremely energetic charged particles, mostly protons, that speed through the emptiness of interstellar space and possibly even the space between the galaxies. Their paths twist and curve as they pass through electromagnetic fields, and they end up bombarding the Earth from all directions.

As they smash the atoms and molecules of our atmosphere, "secondary radiation" is formed, which strikes Earth's surface and is so energetic it passes through anything in its way, including human beings, and buries itself in the Earth.

In passing through the human body, the radiation is bound to slam into molecules and damage them. Usually, this damage is not important and the body can repair it. Every once in a while, however, a cosmic ray may strike a gene and change its structure, producing a "mutation." This mutation can bring about a cancer or some other undesirable condition.

Earth's atmosphere usually manages to absorb and weaken much of the cosmic ray bombardment, rendering it relatively harmless, so that the amount that reaches us is not fatal. Life has continued for billions of years without being noticeably endangered by the cosmic rays.

In fact, some people think the cosmic ray bombardment is essential. The mutations it brings about are generally harmful, but every once in a while, one can be useful. The process of evolution proceeds by the random production of such rare, useful mutations. Without cosmic rays, some think, evolution would be slowed to such an extent that even today there might be no life on Earth more advanced than bacteria.

All this, though, depends on the protective action of the atmosphere. As you go higher, there is less air above you and the number of cosmic rays you receive increases. The people in Denver, Colorado, who are at a high altitude, get considerably more cosmic rays than the people of Los Angeles or New York, who are at sea level. (Nevertheless, the people in Denver are still sufficiently protected.) People in planes that fly at very high altitudes get still more cosmic rays, but they are exposed for merely a few hours.

The difficulty arises, though, when people move beyond the atmosphere altogether and receive no protection at all from cosmic rays. They get them full strength and in full quantity.

Of course, our astronauts have reached the Moon and returned with no ill effects. The shuttle remains in space for a period of time and the people on it suffer no ill effects. Here, however, we are talking of exposures that take, let us say, a week. Astronauts have remained in orbit for longer than that. Some Russian cosmonauts have stayed in space for as long as a year, but have suffered considerable physiological changes in the process.

But what if we talk about a flight to Mars, during which the astronauts will have to remain in space for a year and a half? What if we imagine people working on the Moon over extended periods of time? What if we imagine the building of space settlements, little independent worlds, in which human beings may wish to spend all their lives?

Nor is it only the ordinary quantity of cosmic rays we need worry about. Every once in a while the Sun produces "flares," small explosions on its surface. This produces a burst of energetic radiation, not as powerful as cosmic rays, usually, but powerful enough to damage unprotected human beings in space.

What is to be done?

Obviously, people in space will have to be shielded. Spaceships and space settlements will be enclosed in metal such as aluminum, but the cosmic rays saturate these metals and produce a neutron radiation that can be just as harmful.

Possibly, material from the Moon can be compacted and layered on the outside of space vessels, giving them the protection of rock, though I don't know if that, too, will be saturated by cosmic rays.

One physicist at NASA, Rein Silberberg, is of the opinion that water would be an effective barrier. He visualizes a kind of double wall with water in between. He thinks that a four-inch thickness of water would reduce the cosmic ray radiation to the point where the risk of cancer is increased by only 2 percent over that on Earth, although this may be inadequate in case of a sudden solar flare.

We must solve the cosmic ray problem, as well as other serious consequences of long-term exposure to zero gravity, like the demineralization of bone. Life has always moved on, and human beings have always liked frontiers. We will undoubtedly move out beyond Earth, challenging the new frontier of space.

Invisible "Icy Planets"?

Between the orbits of Mars and Jupiter a belt of asteroids circles the Sun. The first asteroid was discovered in 1801, and since then we have located about sixteen hundred of them.

The astronomer S. Alan Stern suggests that there is a second belt far, far out beyond the orbit of Pluto. In fact, it is likely to be two hundred times as far away as Pluto, or 6 trillion miles, just about one light-year.

The chances of seeing anything that far out with ordinary telescopes is just about zilch. In fact, it might be argued that Pluto and Charon are the largest of these objects and certainly the closest to us and so they can be seen. But that would appear to end it.

What, then, makes Stern and others think that there are invisible "icy planets" a light-year away? For the most part, it is because the outermost planets show such odd behavior. Uranus, for instance, rotates on its side. No other planet has its rotational orbit twisted to such a degree. One hypothesis is that when Uranus was first formed it rotated more or less upright as the other planets do, but early in the course of its formation, it was struck by a rather massive object that twisted it. In fact, it would require an object anywhere from a fifth the mass of the Earth to five times its mass to do the trick. This colliding object may have been one of the "icy planets."

Then there is the case of Triton, Neptune's large satellite. It revolves about Neptune in retrograde fashion. That is, other satellites turn about their planet in the same direction that the planet itself is turning, but Triton turns in the opposite direction. It is the only sizable satellite that does that. The usual hypothesis is that it collided with a massive object early in its history and turned askew. Again an icy planet? Perhaps.

157

And there is Pluto itself, with its satellite, Charon. Pluto is only six times the mass of Charon, closer in size than Earth and the Moon. How did Charon come to be revolving about Pluto?

Pluto might have been struck by an object that broke it in two. Or it might simply have encountered Charon and captured it. Either case is quite unlikely unless there were numerous objects in the vicinity. In fact, there must be numerous objects in the vicinity to suppose that Uranus was struck and turned awry, and that Triton was struck and reversed its orbital motion.

But if there were icy planets near enough to the outer planets to affect them, where did they go?

There's a gravitational pull from the nearer stars. It affects the icy planets at the outer edge of the solar system just as it does the comets (the comets, which are also at that distance, are far more numerous than the icy planets, but also far smaller). Some of the gravitational pull would cause the icy planets to move in toward the inner solar system, where they wouldn't have lasted very long. Others are pulled outward into the vast spaces a light-year away.

Is there any way of detecting the icy planets even if we can't see them? We've sent several probes careening far beyond Pluto, but they have not sensed a thing. Does that mean the icy worlds do not exist? Not at all. Suppose there are three thousand of the icy planets (Stern thinks there may be as many as that). Imagine them distributed evenly over the enormous space. They would be millions of miles apart. Space would seem empty of them, and the chance of an encounter between a probe and one of them would be virtually nonexistent.

If an ordinary telescope can't find them, a powerful infrared one might, for the ice worlds would give off infrared to a far greater extent than visible light.

If the Sun possesses a far distant asteroid belt, other stars might also. There are two stars, Vega and Beta Pictoris, that definitely have dust clouds surrounding them. It is not impossible that part of those dust clouds are the distant ice worlds.

Closer to home, the ice worlds may explain such objects as Pluto, Charon, and the recently discovered Chiron, which is definitely an ice world. It has a cometary makeup but is at least a thousand times the mass of Halley's comet.

The result is that we may discover a great deal more about the origin of the solar system. For that reason, it would be exciting indeed if we could manage to land a probe on Pluto to study it and its satellite closely.

Who knows what we may find out?

Asteroids Around Us

The asteroids are probably the oldest unchanged objects in the solar system. They are so small that they have no oceans, atmospheres, or anything else that can change them, so that they are over 4 billion years old. This means that a close study of the asteroids may tell us a great deal about the early days of the solar system and how it was formed.

What can we find out about these objects that are so small we can only see them as points of light? For one thing, they reflect sunlight, and we can now pick up those reflections with sufficient precision to tell how they vary with time. It is generally believed that these variations exist because the asteroids are revolving (all the heavenly bodies we know revolve) and because they have some surfaces brighter than others. This sort of observation tells us how fast the asteroid is turning. We also find that some asteroids are generally darker than others and that the darker asteroids tend to be farther from the sun. This, too, may give us hints about the early solar system.

There are gaps in the spacing of asteroid orbits, usually called the "Kirkwood gaps," after the astronomer who first noticed them. Partly as a result of this, the asteroids are grouped into families. It is possible that the asteroids formed in the very early days as a result of collisions and that each family represents the breakup of a particular larger asteroid. We don't know.

Asteroids also reflect infrared light, which can be measured. That gives us two pieces of knowledge—the infrared brightness and the ordinary brightness. That, too, helps us decide how large an asteroid might be and how much light it reflects.

What about the shape of asteroids? Every once in a while, an asteroid passes in front of a star and for a few seconds the starlight disappears. Considering the length of time of the disappearance we can judge the width of the asteroid. If enough people observe it at the same time, they can even tell how this width varies with position, and this gives us the shape. It turns out that most asteroids are irregular in shape.

Not all asteroids stay primly and securely between the orbits of Mars and Jupiter. Some wander off beyond Jupiter, and some move in closer than Mars. It is the latter group that is more interesting because these close asteroids can sometimes skim past the Earth at fairly close distances (astronomically speaking).

It is difficult to observe these close asteroids in detail because they move so quickly that it is next to impossible to get one's instruments into action before they have passed out of view. One close asteroid (or "Earth-grazer," as they are sometimes called) is asteroid number 4769. Very rapid work, including studies of previous passages, showed that it is a dumbbell-shaped object.

Nowadays, of course, we don't have to depend on earthbound instruments only. There are probes that, we hope, will pass close by an asteroid and tell us more about it in a few moments than we've learned in all the years we've been studying them from Earth.

The ultimate question about asteroids is whether one will ever strike the Earth. This is quite possible. In fact, if we wait long enough (millions of years, perhaps), it is inevitable. Scientists are increasingly certain that it has happened in the past; such a collision with an asteroid (or perhaps a comet) 65 million years ago may have wiped out the dinosaurs.

There you have a real catastrophe. Another strike of that sort is sure to wipe out civilization and perhaps the human species. That's a good reason for finding out all we can about asteroids so that we can do the necessary calculation to see if some asteroids are coming far too close for comfort. And if one is, what can we do about it? Right now, nothing.

Of course, we need not look upon asteroids simply as a vehicle for catastrophe. Since some do approach us quite closely (within a million miles or so), they would be much more easily reached than, say, Mars, if we ever truly develop space travel. If so, they can be very useful as a source of minerals and metals. Because of the asteroids' insignificant gravity, it would take very little energy to lift those materials, bringing them into Earth's orbit to be used to help build the space cities of which some astronomers and other scientists dream.

Twin Asteroids

On August 9, 1989, a new asteroid was discovered by Elinor Helin of the California Institute of Technology. In itself, this was not startling, for nearly two thousand asteroids are known. On investigation, however, it turned out to be the most startling one we know.

The first asteroid was discovered on January 1, 1800 (the first day of the nineteenth century), by a Sicilian astronomer, Giuseppe Piazzi, entirely by accident. Astronomers had speculated there must be a planet between Mars and Jupiter, but Piazzi wasn't looking for it. He just found it and named it Ceres after the patron goddess of Sicily in ancient times.

Ceres was found to be very small, only about six hundred miles in diameter, much smaller than any other planet, which was why it hadn't been discovered earlier. Other astronomers thought that it was too small and that there must be something else between Mars and Jupiter. By 1807, three other small planets were discovered in that space, all of them even smaller than Ceres. They were so small that even in a telescope they showed up only as points of light, as stars do, and didn't expand into little orbs of light as do other, larger planets. Because of their appearance, the new little planets were called "asteroids," from Greek words meaning "star-like."

As time went on, other asteroids were found and still others, all small, most only a few miles across. They all seemed concentrated in the space between Mars and Jupiter and seemed to be remnants of a planet that had exploded, or, more likely, one that had never formed because the gravitational attraction of Jupiter kept the asteroids from clumping together. The region between Mars and Jupiter where the asteroids were found was called the "asteroid belt."

162

The asteroids in the asteroid belt were anywhere from 40 million to 400 million miles from Earth even at their closest. Because of their distance and their small size, no detail could ever be seen in those bodies.

In 1898, however, an asteroid was discovered by a German astronomer, Gustav Witt, that, in its orbit, moved into the space between Mars and Earth. It was the 433rd asteroid discovered and was named Eros. There were occasional times, with Eros and Earth at the proper points in their orbits, when Eros was only 14 million miles from Earth, closer than any other object but the Moon. In 1931, it approached within 16 million miles, and its distance could be so accurately determined that the distance of all other objects in the solar system was calculated from it. These were the best figures for the size of the solar system till astronomers learned how to bounce radar waves off the planets and calculate distances that way.

Eros is an example of an "Earth-grazer," and, in the last fifty years, numerous additional Earth-grazers have been discovered. About 130 asteroids are now known that actually move in more closely to the Sun than Earth is. Some can, at intervals of time, approach within only a few million miles of Earth. One asteroid, Hermes, which may have been a mile across, missed us by only 200,000 miles in the 1930s, and has never been seen since.

One Earth-grazer or another might possibly collide with Earth at intervals of hundreds of millions of years. A fairly sizable asteroid or comet may have struck Earth 65 million years ago and wiped out the dinosaurs and many other forms of life.

The new asteroid discovered by Elinor Helin is an Earth-grazer. One of the reasons for its discovery was that it was making a close pass at Earth (one it makes only about every fifty years), and it was only 2.5 million miles away.

At that distance, maybe some detail could be made out, especially with today's highly advanced instruments. There is a thousand-foot radio telescope at Arecibo in Puerto Rico, for instance, that is the best of its kind in the world. If a beam of microwaves (like those in radar) are sent out to the asteroid, it will be reflected and the Arecibo telescope can pick up the reflection. A team led by Steven Ostro did this, and from the reflection (just as in the case of reflected light) the asteroid could be "seen."

This was done on August 22, 1989; the reflections were analyzed and to the amazement of all, it showed *two* asteroids that seemed to be in contact, revolving like a propeller once every four hours. No such twin asteroid had ever been seen, though, for all we know, there may be others that we can't see because they are too far away to show details.

Perhaps such a twin asteroid formed because there were two asteroids in nearly the same orbit, flying about the Sun side by side. A tiny gravitational force would pull them together and finally bring them into contact. There they remain in a coy embrace, circling each other, and, every once in a while, skimming by the Earth.

Space Watch

Small asteroids that come comparatively near Earth are called "Earth-grazers." Recently David Rabinowitz at the University of Arizona found the smallest, closest object ever seen outside Earth's atmosphere.

It may have been no more than thirty feet in diameter, and, using sensitive electronic detectors, the asteroid was followed for six hours. The orbit was calculated.

It makes a complete circuit of the Sun in something between three and four years, and at its near point to the Sun, it is as close to it as Venus's orbit is. At the other end, its far point, it recedes to the asteroid belt between the orbits of Mars and Jupiter. In between, it comes near Earth's orbit. And, very occasionally, it passes Earth's orbit when Earth itself happens to be close by. On this occasion it passed within 106,000 miles of Earth, a little less than half the distance to our Moon.

Of course, 106,000 miles is still quite a distance to us, but astronomically, it is nothing. In addition, the orbits of these small asteroids can easily change. For instance, this small one, having approached Earth so closely, must have curved a bit in its path thanks to Earth's gravitational pull, and it may not return in exactly the same way.

A small asteroid that may miss us in its journey by different amounts over the ages may eventually take up an orbit that causes it to zero in and strike us.

Is this something to worry about? Yes, for the impact of even a thirty-foot asteroid could do considerable damage, and it is not merely one asteroid that is under consideration. Astronomers know of fifty asteroids with diameters of one to two kilometers that can, at times, approach us at distances of 20 million miles or less. There may be fifteen hundred of them with diameters of about half a kilometer, and many thousands that, like the one just spotted, are only a few feet across.

If any of these hit us, the results can be catastrophic. Even a strike by an asteroid only thirty feet across, but traveling twenty miles a second, might destroy most of a city if it struck one squarely. An asteroid that is one kilometer across could rip up thousands of square miles. Or, if it landed in the ocean (and the chances are 7 to 3 that it will) it will set up a tidal wave that could drown millions and destroy unimaginable amounts of property along the coastal regions of the world. An asteroid that is ten kilometers, if it strikes, would produce effects that might destroy most of life.

Then why hasn't any of this happened?

Oh, but it has happened. In 1908, some sort of asteroid or comet struck in central Siberia and knocked down every tree for forty miles in every direction. Fortunately, it was an uninhabited area and no one was killed, but a man sixty miles away was knocked out of his chair by the concussion. That was only a small one, maybe like the one astronomers have just spotted.

In Arizona, there is "Meteor Crater," a round crater about half a miles across that was produced when a small asteroid hit about fifty thousand years ago. That had to be larger than the one in Siberia, and if such an asteroid were to strike again in a densely populated region of Earth nowadays, millions would die in a minute.

Most scientists believe that, 65 million years ago, there was a far bigger impact still and that an asteroid or comet struck, one that may have been up to ten kilometers across, and wiped out most of life on the planet, including all the dinosaurs.

Up to now we humans could do nothing about future impacts but wait and hope. While nothing of importance may hit for a million years, there could be a sizable concussion tomorrow. Until now, we couldn't even see anything coming till it was too late, but now, as you see, we can spot fairly small objects at quite a distance.

Clearly, what is necessary is a space watch. There must be satellites or space stations that are forever monitoring near space in the search for such objects. Any object that has the potential of coming too near Earth must be broken up by a nuclear bomb or by some more efficient device that may be discovered in the future. It could in this way be broken up into pebbles. Another solution is to produce an explosion near enough to the object to alter its course and cause it to miss Earth.

Astronomers are now seriously talking about setting up a space watch. Whatever it cost would be made up millions of times over if a single important impact were prevented. I must, however, tell you this. Way back in 1959, in an essay I wrote for a small magazine that no longer exists, I advocated just such a space watch for just the reason I have described here. Astronomers may be considering it now, but I have been advocating it for over thirty years.

More on Meteors

Meteors are the small chunks of matter that zoom in from outer space and heat up in Earth's atmosphere (when visible they're called "shooting stars"). "Meteorites" refer to meteors that have landed, and "meteoroids" are meteors that are still above the atmosphere.

The Greeks knew that "shooting stars" were not actual stars because they could count, and after a meteor "shower," the number of real stars stayed constant. Sometimes ancient peoples saw a big meteor fall and found the actual meteorite. In fact, until humans figured out how to smelt Earth's iron, they used the iron from meteorites. There are rocky meteorites, too, but those were less obviously of unearthly origin.

A few years ago some meteorites found on Antarctica hit the news headlines, for scientists had decided they originated on Mars. Evidence for this is that these meteorites are too young to have come from the very beginning of the solar system, 4.6 billion years ago. They are also too young to have been ejected from the lunar surface, as some other meteorites may have been. The Moon's volcanoes have been inactive for 3 billion years, whereas the volcanic rock of the "Martian" meteorites is estimated to be only 1.3 billion years old. It's probable that a huge asteroid hit Mars with such force that chunks of the planet achieved escape velocity, left Martian gravity, and entered space. Eventually some were captured by Earth's gravitational field. Since the study of the Antarctica specimens began, a few meteorites on other parts of Earth are now thought to be from Mars also.

Analysis of the "Martian" meteorites showed that they contain trapped gases comparable to the Martian atmosphere, already known from the data delivered by the Viking lander. In 1992, scientists presented evidence that the Martian meteorites also contain 0.04 to 0.4 percent water. Oxygen isotopic analysis reveals that the water is not from Earth. Most important, the water was not in "oxygen isotopic equilibrium" with the host meteorite, which indicates that Mars has two distinct oxygen isotopic reservoirs.

Perhaps this means that although Mars once had surface water, the planet is and was very different from Earth. Our planet has plate tectonics, with constant exchange between the ocean water and the stuff coming up from the mantle through the midocean ridges. This recycles the ocean through Earth's crust (taking a million years or more, but what's that in geology?). In contrast, Mars evolved with the rock and the water each doing its own thing instead of profoundly interacting with each other as they do on Earth.

It takes a lot of effort for nature, or humans, to form diamonds from graphite. It was therefore exciting when tiny diamonds were found in meteorites (the first at least a century ago). In recent analysis of the absorption spectra of infrared light from dense molecular clouds in our Milky Way galaxy, scientists have found more tiny diamonds.

It's not known whether or not these diamonds in space are attached to the clouds or free-floating, but the discovery has led scientists to postulate that some of Earth's meteoric diamonds may have been part of the original cloud that coalesced to form our solar system. A few other meteoric diamonds may have formed from the shock of collisions while still in space.

Also in 1992, British and German scientists analyzed another type of meteoric diamond. They concluded that the larger "Abee" diamonds have "typical solar system isotopic compositions for carbon, nitrogen, and xenon." This implies that they were not formed in the galactic cloud that antedated our solar system, but afterward. Furthermore, they seem to have been formed under relatively low pressure, which rules out collision impact. These Abee diamonds clearly need further study, which may reveal more about the formation of the early solar system.

Cosmochemists Paul H. Benoit and D. W. G. Sears have investigated the origins of meteorites. According to analysis, there are two rates of cooling of meteorites, which may mean that one group formed in the parent body at a deeper level than the other. Shallow-formed meteorites may have been broken off and ejected earlier, arriving earlier on Earth. The main conclusion from their work is that the meteor fall on Earth has, over time, varied considerably in type, size, and number.

Other objects falling from space can look like "shooting stars." It's one thing to have meteors fall on Earth, but quite another to have chunks of our own space junk descend upon us. Perhaps the money spent on guns to fight other humans could be spent on cleaning up our own space garbage.

Comet Dust

Up to a few years ago, Earth was supposed to be surrounded by tiny grains of material that couldn't be studied in any way.

But the University of Chicago planetary scientist Edwards Anders has changed matters.

Anders was studying meteorites in order to work out "exotic nuclei." These were nuclei that were the result of the fission of super-heavy elements. As part of his search, Anders soaked the meteorites in strong acids and noticed that there was a tiny residue of fine powders left over. There were diamonds, very tiny ones, plus tiny flecks of carbon and silicon carbide.

These looked like tiny grains that were not part of the solar system, but were formed outside the system billions of years ago and are now landing on the Earth.

Ernst Zinner of Washington University in St. Louis measured the isotopes present in the grains with a tiny instrument called an "ion microprobe." The isotopes can be used to determine the nature of the nuclear furnaces of stars. Or else they can be used to work out the long-dead nature of the stars.

The grains formed in the terrific heat out of which the Sun and the planets came into being. The grains were destroyed under such circumstances, but there were always a few that survived in the cooler outer reaches of the forming solar system. These grains that survived were found in the asteroids that circled the Sun between the orbits of Jupiter and Mars. They were also to be found in comets, and comet dust, which forms constantly, is probably rich in them.

The main thing is that the grains had not formed in the solar system but had originated outside. The nature of the grains and their isotopes could be used to determine the nature of the stars from which they originated.

The fact that diamonds, graphite, and silicon carbide all contain carbon made it appear that the grains were formed from carbon-rich stars. For that matter, the stars would have to have vast atmospheres within which the grains would form. Apparently, the stars in question were red giants.

Anders had to work in general on stars, but then along

came Zinner and his ion microprobe, and he was able to study the isotopes of single grains for information about the star from which they originated.

On the whole, the story of the grains rather agrees with the theory by which elements are supposed to have been formed. However, there are a few items that are puzzling.

Anders points out that by taking the ratio of krypton to selenium the temperature of a star can be worked out. The hotter a star, the less krypton it should have.

There is also some indication that grains can originate from supernova explosions and not simply from red giants.

The existence of the grains also points up the origins of our own solar system. There are grains whose origin seems to have indicated that they were formed near the cloud that formed the solar system and jostled it into action. It may be that the grains that exist in the solar system indicate just how many stars were involved in the formation of the system. There was a time when astronomers felt that five or six stars helped give the solar system cloud a push, but Anders feels that there may be as many as a thousand.

It is also possible to tell how old grains are. As the grains move through space, they periodically encounter a cosmic ray particle. Anders studied the grains to see just exactly what kind of cosmic ray particle struck them. A particle may strike a carbon atom and make it take up a neon particle. As a result, it would seem that the oldest grains would be about a billion years older than the solar system. Some may be even older.

Since comets are undoubtedly very rich in grains and since comets very rarely strike the Earth, the European Space Agency has a plan to aim at a comet that will come fairly close to Earth. It will scoop up some of the icy material and bring it back to Earth for close study. Comets are cold, much colder than asteroids, so they probably have grains in plenty. Unfortunately, it is unlikely that enough money can be scraped up for the purpose. And even if the money could be raised, it could be decades before the project gets off the ground.

From the fact that the grains will tell us about the universe outside the solar system, we could learn a great deal about the universe, and about the formation of our own solar system. Some astronomers suggest that this represents a whole new field of astronomy and that the ion microprobes are the new telescopes.

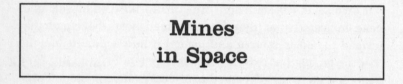

Mines
in Space

If we are going to build a space-centered society that will expand the human range and give us new sources of matter, energy, industrial production, and abstract knowledge, we will have to do so soon. If we wait too long, Earth's increasing and steadily more impoverished population, as well as the planet's deteriorating environment, may make us incapable of the effort.

But if we do build such a space-centered society, we can't rely on Earth's own resources. They are already stretched thin. We must make use of energy and structural materials from regions other than Earth. Energy, obviously, can come directly from the Sun, as we build power stations that convert solar radiation to electricity. But how about the material out of which to build the power stations and the living quarters for thousands of human beings in space?

The nearest source of materials, other than Earth itself, is the Moon. Scientists like the late Gerard O'Neill of Princeton University have, for years, planned the setting up of lunar mining stations. The advantages are that the Moon is quite large, quite close, and has a surface gravity only one-sixth that of Earth.

The disadvantage of the Moon is that it is a "baked" world. Between heat and low gravity it has not held onto materials that evaporate too easily ("volatiles"). This means the Moon does not have the vital elements of hydrogen, carbon, and nitrogen. No matter how we mine the Moon, those elements must be supplied by Earth itself.

The next closest large world is Mars. We know that Mars has a supply of volatiles. We can get everything from Mars that we can get from the Moon, and, in addition, Mars can give us hydrogen, carbon, and nitrogen.

The disadvantages are that Mars is much farther away than the Moon and that its surface gravity is two and a half times as intense as the Moon's. Mars is therefore much harder to reach, and material obtained from it is much harder to lift off its surface.

What next? Well, there are the asteroids. They are small bodies, with negligible surface gravities. What's more, they come in a variety of chemical types. Some are "carbonaceous chondrites," which contain the volatiles we need. Some are metallic, made up chiefly of iron, nickel, and cobalt, and in them these important structural materials are already concentrated, together with smaller amounts of gold and of platinum-type metals. Some are stony and have all kinds of rocky silicate materials interspersed with iron.

Asteroids would be very handy indeed as sources of the materials we need to build a space-centered society, but the disadvantage is that the asteroids are even farther away than Mars, for almost all asteroids circle the Sun between the orbits of Mars and Jupiter.

But notice I say "almost all." There are some asteroids that have managed to find orbits that carry them fairly close to Earth, some even very close. In 1989, one small asteroid whizzed past Earth at a distance of 430,000 miles, less than twice the distance of the Moon.

In the last ten years, more than 125 asteroids have been discovered that are capable of making near-Earth approaches, and there may be as many as a thousand of them altogether. They can be dangerous, too, for they can collide with Earth. Such a collision knocked down trees for forty miles around in central Siberia in 1908, but killed no human beings. A much more serious collision 65 million years ago is thought to have wiped out the dinosaurs.

As long ago as 1959, I wrote an essay in which I suggested that, once human beings developed the necessary technology, a "space watch" be instituted to keep an eye out for such near-Earth satellites, in order to destroy any that might show signs of coming uncomfortably close to our vulnerable planet.

That, however, was a purely destructive suggestion, necessary but not sufficient. Why simply destroy? It would be better to devise ways of capturing these near-Earth asteroids and using them as sources of valuable materials out of which we can build our new society.

These near-Earth satellites sound as though they might be a strictly limited resource, but if we can find even a single nickel-iron asteroid half a mile across and make use of it, we would have a source of metal that would last for many, many years. Besides that, there would always be new near-Earth asteroids being fired at us from the asteroid belt, for the orbits are not fixed but are always being yanked at by the planets, chiefly by giant Jupiter.

Finally, even if after a century we find that we have made use of the best of the near-Earth asteroids and that the supply of good ones is running out, our space-centered society will, by then, surely have advanced to the point where longer trips to the asteroid belt itself will become more or less routine.

There we would have not a thousand, but a hundred thousand objects, some of them several hundred miles across. It would be a resource large enough to enable us to conquer the solar system from end to end, and make it possible for us to fix our gaze on the stars.

III

SCIENCE
AND
TECHNOLOGY

False Alarm

Another "startling scientific finding" has been shown to be a false alarm.

That happens occasionally in science. Something turns up that threatens to change the fundamental view of some aspect of the universe. There is considerable flack and other scientists investigate the matter—and the threat collapses.

The most spectacular case of this sort in the last few years was the sudden excitement over the possibility of cold fusion—the prospect of unlimited energy from very simple equipment after physicists had spent uncounted millions of dollars on huge machines for the purpose without, so far, succeeding.

The cold fusion process was investigated all over the world, and it proved a false alarm. The two scientists who announced it had been premature, and no one else was able to extract significant amounts of energy by the process.

That was not the only case of the sort. There are only four known forces in the universe, and they seem to account for everything that happens. They are gravitation, electromagnetism, strong nuclear, and weak nuclear.

Some time ago, a fifth force was suggested with very strange properties. It was reported as weaker than gravitation, made itself felt for only limited distances of a few yards, and varied for different chemical substances. If this were so, Einstein's general theory of relativity would have been upset. However, careful testing showed it to be probable that no such fifth force exists.

Another time there was a report that tiny bubbles of air trapped in amber showed that millions of years ago, the Earth's atmosphere contained 30 percent oxygen, rather than 20 percent as it does today. This seemed unlikely on the face of it. With 30 percent oxygen, any forest fire started by lightning would sweep a continent. Sure enough, further examination showed that the report was mistaken.

The great supernova of 1987 in the Greater Magellanic Cloud produced a neutron star that spun at a rate of two thousand times a second. So the observations seemed to show. This put astronomers into a considerable tizzy because they could not account for a spin so unusually rapid. But then it turned out that the observations were mistaken. There had been an error, and the neutron star has not actually been seen.

Toward the end of 1989, two Japanese physicists, H. Hayasaka and S. Takeuchi, reported something absolutely astonishing. They set gyroscopes to spinning rapidly and found that under certain circumstances, the faster the gyroscope spun the less it weighed. What it amounted to is that the gyroscope developed some form of anti-gravity. This, in itself, isn't entirely inconceivable. A toy propeller, if set to whirling, will lift itself into the air, as a result of known aerodynamic forces. It wouldn't do it in a vacuum.

In the case of the gyroscope, however, scientists couldn't see what would cause the weight to decrease. Still, if it did, it might be supremely important. It might represent a new way of lifting up into space without firing rockets.

That, however, wasn't the real excitement of the report. The Japanese physicists said that the weight loss took place only if the gyroscope spun in one direction. If it spun in the other, there was no weight loss at all.

This seemed absolutely unbelievable. The universe is run by certain rules of "symmetry." If something happens in one direction, it should happen in another. Thus the Sun would pull at the Earth at a certain distance with the same force, wherever Earth might be, on one side of the Sun or the other or above or below. How could the gyroscope do one thing when spinning in one direction and something else when spinning in the other?

Naturally, physicists all over the world began to fool about with gyroscopes. In February 1990, a particularly careful investigation was reported by T. J. Quinn and A. Picard, two physicists working in France. They ran gyroscopes at speeds of up to eight thousand revolutions per minute under conditions that carefully made certain that they would not be affected by temperature change, by friction, by influences from the outside world, and so on.

They reported that the gyroscopes behaved alike in all respects in whatever direction they spun and that there was no significant weight loss under any condition. Another false alarm.

Does all this mean that no startling scientific finding can possibly be true? Not at all. Every once in a while, some scientist comes up with something totally unexpected and it turns out to be quite accurate. Thus, a couple of years ago, substances were found that were superconductive at unusually high temperatures. This sounded completely unbelievable, but it turned out to be true.

The Shrinking of the Delta

Recent expeditions sponsored by the Smithsonian Institution and the National Geographic Society, under the leadership of the oceanographer Daniel J. Stanley, indicate that one of the most fertile areas on Earth, and one that has contributed to social and intellectual advance for ten millennia, is shrinking.

The region in question is the delta of the Nile River.

Ten thousand years ago, agriculture was invented in the uplands between the modern nations of Iraq and Iran. At first, agriculture depended upon rainwater, which was always chancy, but, very slowly, the farmers moved to nearby riverbanks, where the water supply was much more reliable.

It might seem that rivers were so obvious a source of water that it should not have taken long to move there, but rivers won't work unless you build levees and dig ditches to bring water to the crops and to prevent floods. That required the development of cooperative procedures, and it took time.

Nevertheless, after 4000 B.C., city-states began to rise along the banks of the Euphrates and Tigris rivers (modern Iraq) and along the Nile River (modern Egypt). The Sumerians who settled along the lower course of the Euphrates developed the art of writing by 3500 B.C. and built the first high civilization. The Egyptians along the Nile were not slow to copy the notion, and a second high civilization was built there.

The two were different in this way. The Tigris-Euphrates were open on both sides to nomadic tribes so that the Sumerians not only fought against each other but had to fend off invading Akkadians, Kassites, Aramaeans, and so on. The Nile River, however, flowed through a desert. East and west there were virtually no people, and the inhabitants along the Nile could live in peace and did so for thousands of years, the longest period of stability for any people before or since.

The Nile was a perfect highway, flowing south to north, while the wind blew steadily north to south. By hoisting sails, boats could be driven southward; by lowering the sails the current would carry them northward again. What's more, there were no storms, and the river was always peaceful. Trade was encouraged, and the city-states along the river combined to form a "nation" with its cities sharing a common culture and heritage. Egypt was the first nation on Earth.

Then, too, every year, the melting snows in the east African mountains far to the south of Egypt swelled the Nile and produced an annual flood that put down a layer of fertile silt along its banks. Because it was an annual flood, the Egyptians developed the calendar we still use and they developed geometry so they could lay out the boundaries of the fields again after the flood. They rarely experienced food shortages, for the flood kept the land fertile. In the Bible, when there was a famine in Canaan, the sons of Jacob had to go to Egypt to buy grain— there was always grain in Egypt.

Of course, the fertile strip was just a couple of miles on either side of the Nile River. Toward its mouth, the river spread out into a triangle of many streams, and that was the fertile Nile delta, so-called because it had a triangular shape like the Greek letter "delta." Nowadays all river mouths are called "deltas," even when the shape, like that of the Mississippi delta, is not at all triangular.

Egypt was the richest nation on Earth in ancient times. For a period between 1500 B.C. and 1200 B.C. its armies invaded the upper Nile and western Asia and it formed the "Egyptian Empire." Finally, it fell to the Persians in 525 B.C. and was never truly independent again, but it continued to flourish. Between 300 B.C. and 30 B.C., it was the most advanced nation on Earth, intellectually. It had a great museum, the first organization that could be considered a university, and it had the largest library seen before the invention of printing. And it remained the granary of the Roman Empire.

It continued to be prominent in medieval times as part of the Moslem world. Even today it supports a huge population of 50 million people, all strung along the course of the Nile River. Most of them live and farm in the Nile delta.

But now the delta has fallen on evil times. The Aswan dam, built in 1964 to control the water supply, has cut down on the amount of silt that reaches the delta. The silt already present is slowly sinking, and since not enough silt is coming in to replace that which is sinking, the Mediterranean Sea is advancing inward.

Stanley estimates that in the next hundred years, the sea will advance inward some eighteen miles, and perhaps more if there is global warming and a rise in sea level. It will swallow many square miles of fertile land and be an enormous catastrophe for the Egyptians unless they take measures to protect their coastline.

Garbage!

The word *garbage* may derive from the old French *garbe* for "sheaf." Some people think it's from *garbelage*—removal of refuse. A 1942 dictionary dismisses garbage as "animal or household refuse," or "low or vile things collectively." The *Oxford English Dictionary* adds "worthless or foul literary matter," and a brand-new dictionary includes "incomprehensible data" put into or produced by a computer. A newer dictionary includes as garbage the various nonfunctioning manmade hardware currently orbiting Earth.

Whatever garbage was or is, we have plenty of it. *Garbage* is listed under "Solid Waste" in an encyclopedia, with 360 million tons produced annually in the United States alone. Billions are spent in disposing of the 11 kilograms (5 pounds) each human discards each day. Many think of the huge stockpiles of weapons as useless garbage littering the planet, so there's that disposal problem, too.

Disposal practices have always been inefficient. With the onset of civilization, garbage dumps provided breeding grounds for disease-carrying animals and insects, especially rats and their plague-carrying fleas. Earlier hunter-gatherers were better off, since there weren't so many of them and they left their largely biodegradable garbage behind, to be recycled into the planet's ecology. The key word is *biodegradable*. The garbage produced by industrial societies is generally not at all biodegradable. Work is being done on this problem, but if you engineer a microbe to eat up the plastic and metal in garbage dumps, would it stop there? Would our metal and plastic cities crumble?

Landfills are less unsightly and disease-ridden than open dumps, but overpopulation insures that space on the planet is running out. You can't dump garbage where people don't live (deserts, ice fields, ocean, mountaintops, etc.). On planet Earth everything is connected, so fouling up here louses up there— to say nothing of destroying potentially valuable ecology.

In the meantime, landfills get larger. One of the world's biggest is the Fresh Kills landfill on New York's Staten Island. It's ironic that *Kills* means "death-dealing" in English but in this case comes from the Dutch word for "channel." Someday the city barges will be turned away from Staten Island, unless we really want another Everest looming beside the Verrazano Narrows.

Landfills are problematical. They aren't becoming nice soil overnight. Many plastics and metals stay put almost forever, and even paper can take sixty years to degrade (unless it's in a modern book, which seems to turn yellow and friable in a few weeks). It's stupid to burn the stuff and add to the already dangerous air pollution, but local plans are afoot to build garbage incinerators costing 100 million dollars each that may send out three thousand tons of toxic ash per day.

If dumping and filling have their limits, and incineration is dangerous, then what's left? Better manufacturing is important. Battelle has developed a plastic material called "Biocellat" to make biodegradable memorial candle holders that humans scatter around cemeteries. Refrigeration coolants destroy the ozone layer, so the Sonic Compressor Systems Company hopes to make refrigerators that work by resonating acoustic waves to compress gas. Nondangerous coolants can then be used because there won't be conventional pistons and lubricants that they would damage. These refrigerators will also be cheaper to run.

Recycling garbage should be worked on more, but with care—some chemical companies were recently indicted on charges that they schemed to dispose of toxic heavy metals by mixing them with fertilizer for sale.

Exporting garbage is difficult, as in the case of the garbage scow that couldn't find a port to accept its load. Isaac once wrote a short story about the scorn aliens had for Terrans who disposed of their garbage on Earth instead of sending it off the planet. The back side of the Moon is a big place . . .

Then there's the problem of garbage we already have in space. NASA is considering heavy shielding its proposed space station to protect it from encounters with junk left in orbit— thirty thousand pieces of it small enough to act like bullets able to penetrate station walls and the space suits of astronauts doing space walks. Even a fleck of paint pitted a shuttle's windshield so badly it had to be replaced. In the foolishness of the arms race, many spy satellites and rockets were accidentally or deliberately exploded in space, leaving fragments in orbit.

The one-in-thirty chance of a space shuttle colliding with space junk has no doubt gotten worse. An engineer at Johnson Space Center has patented a sort of space broom to sweep up orbiting debris, and the work is on to make it do this selectively, so useful things will remain in orbit.

While we worry about the garbage piling up around us and over us, let's remember archeology, a science that isn't just finding beautiful cave paintings or interesting standing stones. Much of archeology is rummaging through ancient garbage, and archeologists love it. They search through the garbage of many Troys to find which one was Homer's. Right now they're exploring the actual living conditions and creative efforts of black slaves in the seventeenth and eighteenth centuries.

Someday archeologists visiting from *elsewhere* may chortle with glee at all the artifacts to examine on planet Earth. Will we have garbaged ourselves into extinction by then?

The
Critters Within

"An increasing opportunistic infection"; "possible origin of common ailment found"; "mosquito-borne infections gaining." These headlines refer to different diseases, yet all are caused by animals that live inside people. Parasites (creatures able to exist only if they live in or on other life forms) can be plants, bacteria, viruses, and so on, but in medical terminology the word is restricted to animals that parasitize humans.

From the medical point of view, history is a record of human folly about sanitation, food, and the places chosen for working and living. Now, according to recent reports, intestinal worms caused twenty thousand deaths last year; hookworm caused sixty thousand; snail fever (schistosomiasis) caused 200 million reported cases and probably many others not reported; and river blindness (caused by a filarial worm called *Onchocerca volvulus*) affects as many as 20 million people in Africa and South America. There are many other examples of how parasites are still affecting humanity.

Parasitic diseases can be fought with education, adequate diagnosis, and new drugs. Recent research, mainly in the United States and Great Britain, has even attacked that nasty insect the mosquito, which brings us more kinds of parasitic and viral diseases than any other "vector."

A recent workshop on mosquito molecular biology and genetics discussed ways to control mosquitoes. Genome mapping of certain strains of mosquitoes is being done to find the genes that seem to make them less able to transmit diseases. Scientists then hope to clone the disease-resistant ones. Ways will be found to help these disease-resistant mosquitoes take the place of infected mosquitoes in the environment.

In addition to other diseases, mosquitoes transmit that dread scourge malaria, which affects people living in more than a hundred countries. The malarial parasite is a protozoan living primarily in red blood cells. It kills more than a million people each year. Right now, not only is malaria increasing in scope, but the usual drugs to combat it are losing effectiveness.

Many kinds of parasites can cause a few symptoms in otherwise healthy humans, but most cause obvious disease, which can be deadly. There are too many parasitical diseases to cover in a short article, but a few have become newsworthy.

Some children thought to have emotional problems have become well when their pinworm infection was cleared up. Ordinarily, pinworms don't cause much more than an anal itch, but apparently children can be more adversely affected, although no one's quite sure why. Small children are more likely to be infected because they aren't careful about what they put in their mouths.

Then there's "irritable bowel syndrome," usually attributed to one's heredity or one's emotions. At a recent meeting of the American College of Gastroenterology, Dr. Leo Galland said that in many cases of irritable bowel syndrome he's used a new test that finds infestations of the trophozoite *Giardia lamblia* in the intestines. *Giardia* is everywhere, and was always thought to be innocuous, unless the patient's intestine was heavily infested.

Parasitical infections that ordinarily give few symptoms are much worse in immunocompromised patients. For instance, lately there have been more serious, sometimes fatal infections from *Strongyloides stercoralis*, a threadworm picked up from soil the way hookworm is. The problem is made worse by the fact that most parasites can suppress the immune system even in healthy people.

On to parasites you get only by eating raw food. Are you still with me? It's probable that before the invention of fire humans were more heavily parasitized than after, for cooking food destroys many parasites. The list of parasites transmitted through raw food is long, but a few are worth noting, in view of current trends in food habits.

Diphyllobothrium belongs to the genus of *Cestodes*, or tapeworms. Its name comes from the Greek words for "two," "leaf," and "little ditch." The most important *diphyllobothrium* is *D. latum*, the fish tapeworm, with an interesting life cycle. Its eggs, from human feces, are eaten by freshwater fleas. Inside the flea the egg develops into a form called a "procercoid." If a fish eats the flea, the procercoid develops into the plerocercoid form in the fish's muscles. It's still so small it can't be seen when the fish is examined. If eaten by a human, it stays in the intestine and grows to the adult tapeworm, which promptly starts producing eggs in four to six weeks, starting the cycle all over.

D. latum is found mostly in northern areas of the world, especially where people eat raw fish. Under the microscope, its spatulate, grooved head is oddly amusing, but it is not a funny disease since *D. latum* is the largest human tapeworm and gives people intestinal difficulties and even severe anemia (because the worm eats all the vitamin B12 and folic acid, depriving the human host). The disease is diagnosed by examining a fresh, wet sample of feces for eggs. There's an adequate treatment, but the most sensible thing is to cook all fish, even those supposedly taken only from saltwater. Freezing fish at $-10°C$ for at least two days will supposedly kill the parasite, but I'll stick to cooked fish, thank you.

There's a pork tapeworm, but most people know enough to eat only cooked pork, because of another, worldwide parasite—*Trichinella spiralis*—which causes the severe disease of trichinosis and can't be detected in uncooked pork. Most good butchers don't grind pork in the same machine as that used for other meats, which might not be so thoroughly cooked.

Which brings me to steak tartare and beef tapeworm. Enough said.

Monsters

The best-known monster of today is the Loch Ness monster, sometimes referred to, more or less affectionately, as Nessie. Loch Ness is a long narrow lake in Scotland, and Nessie is considered to be something like an extinct plesiosaur with a long neck, a long tail and a large body.

No one has ever actually seen Nessie, and in my opinion no one ever will because it doesn't exist. Loch Ness is simply not big enough to house a plesiosaur; it is unimaginable what it could possibly live on; and despite all kinds of investigation, no one has ever spotted it.

Then, why does it remain so popular, and why do so many people think it exists? In the first place, people are always pleased when know-it-all scientists prove to be wrong. In the second place, the locals make large sums out of the tourist trade, and people visit Loch Ness chiefly in the hope of spotting Nessie.

Actually, though, mankind has lived with monsters, usually far more fearful than poor Nessie, throughout its history. The fact dates back, no doubt, to the time when the early ancestors of man moved about in constant fear of the large predators about them. Fearful as the mammoths, sabertooths, and cavebears may have been, it is the essence of the human mind that still worse could be imagined. The dread forces of nature were visualized as superanimals. The Scandinavians imagined the Sun and Moon to be pursued forever by gigantic wolves. When these caught up with their prey, eclipses took place.

Relatively harmless animals could be magnified into terrors. The octopi and squids, with their writhing tentacles, were elaborated into the deadly Hydra, the many-headed snake destroyed by Hercules; into Medusa with her snaky hair; and into Scylla with her six heads.

Perhaps the most feared animal was the snake. Slithering unseen through the underbrush, it came upon its victim unawares. Its lidless eyes, its cold and malignant stare, its sudden strike all served to terrorize human beings. It is no wonder that the snake is often used as the epitome of evil—as in the Garden of Eden.

But imagination can improve even on the snake. Snakes can be imagined who kill not by a bite but merely by a look, and this is the "basilisk."

Or else make the snake much larger, into what the Greeks called "python," which represented the original chaos that had to be destroyed by a god before the orderly universe could be created.

A Greek word for a large snake was *drakon*, which became the most popular of all monsters—the "dragon." To the snaky length of the dragon were added the thicker body and stubby legs of that other dreaded reptile, the crocodile. That gives us the monster Tiamat, which the Babylonian god Marduk had to destroy in order to organize the universe. Think of the burning bite of the venomous snake, and you have the dragon breathing fire.

Some monsters are, of course, animals that have been misunderstood into beauty rather than horror. The one-horned rhinoceros may have contributed to the myth of the unicorn. The ugly sea cow, with its flippered tail, rising half out of the sea and holding a newborn calf to its breast, may well have been transmogrified into a beautiful mermaid.

Throughout history, man's greatest enemy was man himself, so it is not surprising that men served as the bases for some of the most fearful monsters—the giants and cannibalistic ogres of all sorts. Some of these stories arose when primitive tribes encountered much more advanced civilizations. Thus, the primitive Israelite tribes, on encountering walled cities and well-armed soldiers, felt the Canaanites to be a race of giants. Traces of that belief remain in the Bible.

Then, too, a high civilization may fall, and those who follow may forget the civilization but be aware of the gigantic ruins it left behind and feel that they could only have been built by giants. The primitive Greeks, coming across the huge thick walls that encircled the ruined cities of the earlier, more civilized Myceneans imagined these walls to have been built by giant one-eyed "Cyclopes."

Such Cyclopes were later placed in Sicily. They may have been sky gods, and their single eyes may have represented the Sun in heaven. It may also have arisen from the fact that elephants roamed Sicily in prehistoric times. The skull of such elephants, occasionally found, would show a large nasal opening in front which could be interpreted as a single eye.

As man's knowledge of the world expanded, the room available for the dread or beautiful monsters he had invented shrank and belief in them faded. It's a loss in a way.

Noise

Most people say they hate noise, even those who become restless in a particularly quiet place. The older generation complains about the noise young people make, forgetting that their parents probably made the same complaint about them.

Dictionaries usually define *noise* as harsh, loud, discordant sound. Although noise is more complicated than that, *discordant* is a good word, for that's what scientists mean when they say noise is a random and therefore unpredictable signal.

Noise pollution is ordinarily man-made and can literally endanger us. Stand near an airplane taking off, or a jackhammer tearing up the street, or a superamplified rock band. Do it too long, and you may permanently damage your hearing—and more.

In Bulgaria, researchers found that continuous loud noise caused a dramatic drop in the fertility of laboratory animals, whose few babies were smaller than normal. A human baby's birth weight can be lowered when its mother is exposed to excessive noise. After birth, infants and children learn slower in noisy environments. Noise-battered adults can experience hearing loss, headaches, high blood pressure, increased heart disease, fatigue, dizziness, lowered immune functioning, cognitive impairment, and emotional strain.

Medical researchers have shown that healthy volunteers coped better with noise they could control. Lack of control over noise caused severe alterations in mood as well as detrimental changes in the autonomic nervous system. If your next-door neighbor is playing a stereo at three in the morning or running a lawn mower when you are taking your nap, the sound may not damage your hearing, but its uncontrollability will be unusually irritating.

The hearing damage done by regular attendance at rock discos and heavy metal concerts is well known. It's especially bad when the music lover has a noisy daytime job. Some intelligent rock fans are striving to have the volume of sound lowered, and many musicians now wear earplugs. Even "moderate" noise can affect the voltage produced by individual brain neurons in unpredictable, paradoxical ways. No wonder people feel confused in situations of loud noise.

The trouble is that not all bad noise is loud. Low-frequency sounds similar to those produced by storms or much human machinery can cause dizziness and nausea, as well as adversely affect a person's ability to think creatively. Inaudible "infrasound" can make people feel as if their chests were vibrating and their eardrums flopping back and forth.

Electronic noise, not necessarily audible to us, bollixes up our computers and communication systems, including our TV sets. This can be more than annoying. It is positively dangerous when the affected electronic systems are computers running our banks, government offices, military establishments, and—you name it, for computers are everywhere.

An enormous amount of effort is currently put into noise control. Machines are mounted on vibration-absorbing platforms, not only to reduce the noise for human ears, but to prolong the life of the machine. Soundproofing is big business. A new technique of "Active Structural Acoustic Control" is being studied for use in airplane fuselages, marine hulls, and various industrial applications. A structure's modal shapes are changed enough to diminish the radiated sound. If the technique proves as effective as it seems to be so far, it will save everyone money and reduce noise pollution.

"Noise" is, again, how you define it. A tourist in Africa may think elephants are noisy. They trumpet, shriek, and have stomachs that constantly rumble. Unknown to the tourist—unless he feels the vibration—the elephant is also vocalizing at a frequency below human ear capacity. Nobody knows whether or not this is a form of genuine communication. Let's hope we find out before all the elephants are killed for ivory.

Many animals make and respond to sounds we cannot. Small animals hear and produce higher sound frequencies than we can hear. To kill or drive away house mice and cockroaches, we buy little gadgets that produce what are to us ultrasonic waves. Good "noise" for us, bad for the pests.

Intentional noise can be annoying but is helpful, like the beeping of a truck when it backs up. Researchers have found that the murmuring, clucking, clicking noises parents make do indeed calm and attract infants. Apparently all mammals make clicking noises to their offspring!

Noise is also used to confuse spies in the opposite camp, whether that camp be military, political, or economic. Information is transmitted in secret by using a "semichaotic" system (don't ask us to explain this, for we don't yet understand chaos theory, much less semichaos).

And finally, Northwestern University researchers have discovered that the sound-wave map of a primate warning cry is similar to that of a sound that especially bothers humans—the screech of metal or fingernails being scraped across slate. Maybe rock bands are unconsciously screeching to ward off predators?

Cooling Down

One sure way of getting energy is to obtain it from the Earth itself. There are regions of hot springs, and it is only necessary, then, to dig holes in the ground and let the heat come up and be converted to electricity.

One such region, 115 miles north of San Francisco, began to be exploited in 1960. It seemed to offer cheap, smog-free electricity from the Earth's interior, and there seemed to be more and more of such electricity produced. By 1990, it was suggested that two thousand megawatts of electricity would be produced.

Unfortunately, the amount of electricity being produced is not two thousand megawatts but only fifteen hundred. What's more, the steam pressure in the ground is plummeting. The trouble is that, though there is plenty of heat in the ground, there is *not* plenty of water.

What happened was that too many engineering devices were put into use. All one had to do was to drill a bunch of holes down into the steam-filled cracks in the rock and wait for heat and electricity to come out. And it did, too, but as more and more holes were drilled, less and less heat and electricity emerged. By the late 1980s, it was clear that the field was overdeveloped.

Meanwhile, though, the engineers were convinced that enough steam would be developed to produce three thousand megawatts of energy—enough power for 3 million people. They felt that this could be carried on for thirty years. But they were all wrong.

If the region had been developed slowly, then there is good reason to suppose that it would continue in reasonable fashion. However, that's not how it worked. Oil prices skyrocketed during the oil embargo, and it was clear that geothermal energy could be siphoned off. Then, too, the government, at that time, provided economic incentives to make sure that geothermal energy would be developed. The result was that development accelerated, jumping from 70 megawatts to 150 megawatts per year. By 1988, generating capacity had more than doubled from the 1981 level. The new engineering levels kept on going, even though it was clear that steam pressures were dropping at an accelerating level.

By now it is clear to everyone that there isn't enough water in the rocks and crevices below the field to keep things going, but development continues. In fact, the California Energy Commission has approved every step that served to continue to develop the field. Experts are trying to determine methods for rejuvenating the field. The best chance would seem to be injecting cold water into the field, using steam that is condensed before it vanishes into the atmosphere. Even so, water is so scarce that operators are likely to have to pipe in treated sewage to recharge the field.

Other regions of hot springs must be exploited cautiously. Heat near the surface of volcanic areas of the United States is expected to be about ten times the heat energy of all U.S. coal deposits. The catch is, of course, to supply water. Engineers drill into the dry, hot rock and fracture it. They then send down water, which comes up again through another drill hole. Flow tests will check to see whether the system can maintain its heat output without cooling off too fast or leaking excessively.

Geothermal energy arises not from mere hot rock, but from magma—molten rock that lies five to seven kilometers down. No hardware durable enough to work its way through such magma is available, although people drill in Long Valley, east of Yosemite National Park—trying to reach six kilometers down, or a temperature of 500°C, whichever comes first.

Geothermal energy is also available in Texas and Louisiana, from a hybrid of geothermal energy and fossil fuel reservoirs. The deposits formed between 15 million and 18 million years ago, when seawater was trapped in porous beds of sandstone between impermeable clay layers. As more sediments piled on, the material became pressurized. In addition, methane released by the decay of organic matter formed.

In the 1970s, there was some speculation that a simple drill hole into a geopressured zone would unleash a gusher of geothermal heat plus a quantity of methane, or natural gas. Ten years of research have reduced optimism, for so far, experience with ordinary geothermal energy has made people uncertain about it as a new energy source. More work is necessary.

Making Hydrogen

A recent World Energy Conference dealt chiefly with the prospects of making hydrogen, as a fuel, because the burning of fossil fuels (wood, coal, oil, and natural gas) may be seriously affecting Earth's climate. All contain carbon atoms, which in burning combine with oxygen to form carbon dioxide, and the hydrogen atoms combine with oxygen to form water.

Carbon dioxide tends to retain heat, and as we burn fuel, we pump millions of tons of carbon dioxide into the air, so the Earth experiences a warming trend that is called the "greenhouse effect." Since this might eventually become catastrophic, we should stop burning carbon atoms and concentrate on hydrogen atoms. The formation of water does us no harm.

However, hydrogen does not occur as such on Earth. The hydrogen that exists is combined with other types of atoms, and it must be forced away from such a combination. The easiest way of doing this is to break up the carbon-hydrogen combinations in natural gas and store the hydrogen. In doing so, however, carbon dioxide is formed, which is what we don't want.

One way of getting hydrogen without producing carbon

dioxide is to start with water, which consists of hydrogen-oxygen combinations. If this combination is forced apart, the hydrogen can be stored, and oxygen can be released into the atmosphere, where it will do no harm. Besides, when the hydrogen is burned, it will combine with the oxygen once more and form water again.

To separate water into hydrogen and oxygen, we have to run an electric current through it, a process called "electrolysis." However, an electric current is a form of energy that is formed most cheaply and easily by burning coal, oil, or gas, and that produces carbon dioxide, which, again, is what we don't want.

We must therefore produce an electric current by using some form of energy that doesn't involve burning fuel. For the purpose, we can use falling water (as at Niagara Falls), or windpower, or nuclear fission. Falling water is restricted to certain places, wind is erratic, and nuclear fission is feared by the public. A logical alternative is the use of direct solar power.

Sunlight falling on "photoelectric cells" will produce an electric current, which can then be used to electrolyze water and produce hydrogen. Photoelectric cells are expensive, however, and are usually not efficient. That means they cannot compete with the burning of fuel as a source of energy.

Every effort must be made, then, to make photoelectric cells cheaper and more efficient. The best hope would seem to be the use of silicon, which is the second most common element in the Earth's crust—we won't run out of it. But silicon atoms only exist in combination with other atoms, and it is rather expensive to try to break up the combinations to prepare silicon in the pure form needed for photoelectric cells. For that reason, the less silicon we can use in photoelectric cells, the cheaper they will be.

In the past, silicon has been formed as crystals in which the atoms are arranged in very orderly fashion. Silicon crystals can convert 30 percent of the energy of the sunlight falling upon them into electricity, which is fairly good efficiency. However, making the crystals is tedious work, and this adds to the expense.

John Ogden and Robert Williams of Princeton University work on "amorphous silicon," a form in which the atoms are arranged every which way. Amorphous silicon is less efficient than crystalline silicon, converting into electricity only 6 to 13 percent of the energy of the sunlight that falls upon them. However, amorphous silicon is much easier to form than crystalline silicon is.

Furthermore, it would be helpful if, instead of making use of bulky crystals, one were simply to use a thin film of silicon atoms sprayed onto glass or plastic. The weight of silicon film used to make a photoelectric cell would be as little as 1/200th of the weight of silicon crystals that would have to be used. What's more, Ogden and Williams point out that it is easier to prepare a thin film of amorphous silicon than of crystalline silicon and that film-formation might be automated.

Despite the lesser efficiency of amorphous silicon, then, this means it could be cheaper to produce a given amount of electricity by the use of photoelectric cells made up of amorphous silicon films than by any other kinds of photoelectric cell currently known.

We may look forward, then, to the possibility of large arrays of silicon-film photoelectric cells in sunny areas, producing vast quantities of electric current out of sunlight. This current will be used to electrolyze water and produce hydrogen. That hydrogen can then serve as the clean, nonpolluting fuel of the future.

The First Step
in Synthesizing Life

Three scientists at MIT, Julius Rebek, Tjama Tjivikua, and Pablo Ballester, recently produced a synthetic molecule that shows a key property characteristic of living systems and that may offer a hint as to how life came to originate on Earth.

All forms of life can multiply themselves, or "reproduce." Human beings have children; other life forms, from the largest to the smallest, also produce offspring. Even organisms consisting of but a single cell can divide in two, making two cells where there was once one. This is true of all organisms right down to the smallest of all, the bacteria.

What makes this possible is that all life forms contain certain nucleic acids, usually symbolized as DNA and RNA, the former being the more fundamental, which direct the formation of the enzymes that keep all cell chemistry working and all organisms alive.

The nucleic acids have the capacity of "replication." They are made up of long chains of units, called "nucleotides." A nucleic acid can attract individual nucleotides in the cell fluid, line them up in precise order so that they combine to form a second nucleic acid just like the first. In this way there is a continuing supply of new nucleic acids for new cells and for new organisms. Nucleic acid replication is the key process in the formation of all living organisms.

The replicating nucleic acid is found in small structures called "chromosomes," which exist in the nucleus of the cell. There are some life forms called "viruses" that are far smaller than bacteria and seem to be nothing more than chromosomes on the loose. Once a virus has penetrated a cell, it can replicate itself repeatedly.

But how did all this start?

Biologists are quite satisfied that living organisms have evolved slowly over the course of the Earth's 4.5-billion-year lifetime. About a billion years after Earth was formed, the first life forms came into existence as bacterialike cells, and for 2 billion years that was all that existed on the planet. Then more complicated cells evolved, followed by "multicellular" organisms. We know this by vast and detailed studies not only of presently existing life forms but of the "fossils" left behind by organisms long dead.

All those organisms, however, from the very smallest, had nucleic acids. The nucleic acid molecules in simple organisms are more or less as complicated as those in the most complex, consisting of chains of up to thousands of nucleotides. How, then, did the first nucleic acid molecule form?

We don't really know. What makes it worse is that in order for nucleic acids to undergo replication, they must make use of enzymes—but the enzymes only exist because the nucleic acids direct their formation. Neither one can do its work in living tissue without the other, so which came first?

Nucleic acids must have evolved in the first billion years of Earth's existence, but no trace of that process remains.

Here is what *may* have happened. The tiny, very common molecules that existed in Earth's youth—water, methane, ammonia, carbon dioxide, and so on—combined with each other (making use of the energy of sunlight, or of volcanic heat) to form larger, more complicated molecules. Eventually, one such molecule developed that had the ability to undergo replication without the help of enzymes. As it replicated itself, perhaps clumsily, it sometimes made slightly more complex versions of itself that could replicate more efficiently, and that could even direct the formation of enzymes. Those enzymes might then cooperate in the process, making replication far more efficient, and the first, simplest forms of life thus came into existence.

But what was this first very simple compound that could undergo replication without the help of enzymes? It may not exist on Earth anymore, but surely chemists might make it in the laboratory.

Rebek and his team have put together a compound called "amino adenosine triacid ester," which seems to have that property. It is composed of two parts. One part seems to be a piece of a nucleotide (call it A); the second part is something not related (call it B). The two can be put together so that it is A-B.

If the A-B is surrounded by individual A's and B's, the A portion of A-B attracts a B, while the B portion attracts an A. The attracted B and A attach themselves to the A-B, then hook together to form a B-A. The two parts of the combined molecule then separate, and you have two A-B's. Each one can then undergo the process again, and in the end you can fill the solution with myriads of A-B molecules.

This is the first time that the ability to undergo replication has been found in a molecule much simpler than a nucleic acid. The process is slow and clumsy compared with the blinding speed with which nucleic acids work, but this simple compound, amino adenosine triacid ester, doesn't need enzymes to make it work. Did a compound something like this appear on Earth long ago, and did it eventually evolve into nucleic acids? Maybe!

More Replication

Replication isn't a beautiful word, but in action it makes all the beauty of life possible. Replication can be defined as "making a copy," "repetition," or "making a reply."

In natural replication, large multicellular animals do not make exact copies of themselves. Their replication occurs as a product of sexual reproduction, so the resulting offspring is unique. Other animals (notably many insects) can make an exact copy, or clone—an organism formed nonsexually from one parent.

These two ways of replication actually occur on the cellular level. In asexual "mitosis," the DNA-containing chromosomes double; then the cell divides so that each daughter cell is exactly like the original.

When the chromosomes of a cell don't double, the cell divides with a "haploid" complement of genetic material. This is called "meiosis" (from a Greek word meaning "to make less"). As a sperm fertilizes an egg, the double helix is restored, with a new mixture of genes.

Unexpected changes can occur in either meiosis or mitosis. The genes themselves can mutate, or the chromosomes may not split properly. Scientists are currently working on the biochemistry of cell division. For instance, they've discovered a "cell division cycle protein" (cdc2), which, with the help of another protein called "cyclin," plays a key role in mitosis. Changes occur in the linked proteins that line the membrane around the cell's nucleus, presumably so the cell can then divide. Since cdc2 protein occurs in all cells (human cdc2 protein is said to be 63 percent similar to that of the most primitive yeast!), it probably goes back to the early days of cellular life on earth.

Other proteins controlling the various changes in the cell cycle are being investigated and may have important applications in the research of cancer, where cell cycles go awry.

All natural cell division is possible through the self-replication of nucleic acids in chromosomes. When Julius Rebek, Jr., and two other organic chemists produced a self-replicating molecule simpler than nucleic acid, it made the news because its action may resemble primordial Earth's first self-replicating molecules.

More recently scientists in Zurich and Strasbourg constructed synthetic micelles—electrically charged colloidal particles of polymeric molecules. A polymer is a combination of many small molecules to make one large, complex molecule —like starch, proteins, and (in human factories) nylon. The newly constructed micelles replicated when chemical reactions produced more compounds to form their membranes. The scientists speculate that life is essentially a closed process, separated from the environment by a boundary (like a membrane), replicating itself from within its own boundaries.

Rebek (with Jong-In Hong, Qing Feng, and Vincent Rotello) has gone on to develop two closely related, synthetic replicating molecules that coexist and help each other's replication. Exposure of one of the molecules to ultraviolet light caused a "mutation" that enabled it to replicate more often. The mutated molecule could outdo the nonirradiated molecule in taking from its environment the raw materials for replication. Soon the mutated molecule eclipsed the other, similar to an evolutionary "survival of the fittest."

These synthetic molecules are not those that predated life as we know it on Earth, but they express, as Rebek says, "the same behavior (replication, cooperativity, and mutation) as those that were."

In the meantime, another sort of synthetic "life" goes on in computer research. Programs have been devised that can evolve independently to solve problems. Then there are self-replicating computer programs that start spreading and, as secret "viruses," can bollix up legitimate programs. Lately, computer scientists have found a way to help self-replicating programs replicate more efficiently. Quite a change from the days of the first computers, which didn't even have a hard disk to get virus-ridden.

The change in ideas of replication, both natural and synthetic (chemical or computer), is mind-boggling. In 1992, Santa Fe, New Mexico, hosted a conference on artificial life. There was a day for "hardware" (if you don't know what that means, you haven't ever seen a computer or even a pocket calculator). Then there was a day for "software" (you know what that is if you've ever used a computer with a program inserted into it, which has to be done in order to make the hardware do the work you want it to do, like word processing).

Most amazing, the conference also had a day for "wetware," the synthetic replicating molecules nurtured by the likes of Dr. Rebek. There was even an artificial life "4-H Show" (as in rural livestock exhibits), during which scientists showed their favorite forms of artificial life.

Someday soon, perhaps, scientists of all these persuasions will combine synthetic chemical "life" with computer "life" to produce a living being that can evolve. Will this be the ultimate "reply" our human life gives to the silence of the universe?

Nanomagic

Nanotechnology is doing things on a very small scale, for it's down at the level of molecules and even atoms.

Ultraminiaturization started in the electronics industry and has expanded enormously. Microminiaturized hardware is now being created in labs and workshops like the National Nanofabrication Facility at Cornell University and many other places in several countries. Those of us who are still astonished by the transition from radio tubes to transistors are reeling from the very notion of devices so small that their components are measured in nanometers.

So far, the smallest microminiaturized component is about fifty nanometers, which is the width of five hundred hydrogen atoms side by side. One nanometer equals ten to the minus nine meters equals ten angstroms. Since an angstrom equals one hundredth-millionth of a centimeter, that's small.

Obviously, nanostructures are a lot more than itty-bitty silicon discs. Workers in the field talk about quantum wells (this has something to do with electrons confined in two dimensions by ultrathin semiconductor films) and quantum wires (smaller, with one dimension). The idea is that devices using these "quantum" thingamajigs can operate with less expenditure and waste of energy. They already do operate in satellite microwave receivers, fiber-optic communications systems, and even in compact disc players.

The methods for making nanostructures are as fascinating and difficult to understand as the structures themselves. For instance, there's the scanning tunneling microscope, invented only eleven years ago. This microscope shows how individual atoms are arranged and can also be used to mark surfaces.

Other exotic techniques permit the carving out of tiny hills and valleys—on the molecular level—or building up nanostructures atom by atom. On a very human level, these techniques may result in such practical devices as microscopic Velcro for surgeons to use, particularly in delicate areas—like brain surgery.

It's mind-boggling to think that as the twentieth century comes to a close, there are people out there moving atoms around, and not just for the fun of it. The future may depend on their work. Right now researchers are forging ahead on the problem of pushing down the nanometer limit from fifty to ten, the wavelength of electron waves. If they succeed in making a "quantum interference transistor," they'd have a microminiaturized device that would use and waste even less energy.

As data storage gets more microminiaturized by the week, scientists (mainly in Japan) are seriously considering the feasibility of someday devising the multiple microminiaturized components and connections that characterize the human brain (which, let us hasten to add, are not themselves completely understood). Can it be that someday Isaac's old concept of a "positronic" robot brain will be a reality? We can hear all the "Trekkies" out there yelling that Mr. Data already has one.

But why laboriously construct these nanostructures? Why not let them assemble themselves, the way those microminiature devices called "living cell components" do? That's being studied and the work has begun.

Theoretician K. Eric Drexler has worked on and promoted nanotechnology for years. He envisions many fascinating developments, including another that physicians would welcome—molecular machines that could repair defective or diseased human cells. Fantastic voyage, anyone?

Nanotechnology, on its way to becoming big business, will make an even bigger difference not just to the world of research but to the planet's economy and politics. We must save energy. We must become a more global society, possible through better communication. It will be a different world, but surely no one should seriously complain—these days—about that. And if nanotechnology is considerably harder to understand than radio tubes, just use the little words we say to ourselves to make the whole thing clear. They are the words told to television's Maxwell Smart, when he was puzzled by the idea of microminiaturization:

"Eensy-weensy, Max."

That it is.

Fantastic Fullerenes

Buckminster Fuller, who invented the geodesic dome, had a charismatic personality. When he gave a lecture (I heard one once), it was mentally stimulating, even if difficult to understand. Now a molecule named in his honor is stimulating scientists—and the media. Trying to write a very short article about that molecule is a little like trying to explain on a three-by-five card exactly how the Roman empire collapsed. Gibbon took considerably more space to do that.

To begin with, a molecule is a group of atoms. If the atoms are not all the same, the molecule is called a "chemical compound." Some are simple, like water, and some are complex, like most organic compounds. Chemists had a hard time deciphering the structure of some of the larger molecules. One well-known breakthrough occurred in 1865, when German chemist Friedrich August Kekule von Stradonitz fell asleep on a bus and dreamed that a chain of carbon atoms turned into a snake that coiled around to grab its own head. Kekule awoke certain that in benzene the atoms formed a hexagonal ring structure—and he was right.

In 1984 there was a different kind of breakthrough. At Rice University in Houston, Richard Smalley, Bob Curl, and British chemist Harry Kroto wanted to reproduce in the laboratory the way that chains of carbon atoms seemed to be made in giant carbon stars. When they hit graphite with laser beams, they found that large, stable carbon molecules were formed, different from any type of carbon yet seen.

This was called "carbon 60," because sixty carbon atoms arrange themselves in hexagons and pentagons to form a shape remarkably reminiscent of a soccer ball. Or of a Buckminster

Fuller geodesic dome—hence the name "Buckminsterfuller-ene," or "buckyball." This is the giant named "Molecule of the Year" by *Science* magazine.

After studying the results of Smalley's lab, physicists Donald Huffman and Wolfgang Kratschmer discovered that back in 1982 they had accidentally made buckminsterfullerene when they electronically evaporated graphite enclosed in inert gas. This technique proved to be a way of producing large quantities of fullerene. Other techniques have been found, so fullerene research is going on, with scientists discovering more and more about the new molecule and starting a whole chemical family of fullerenes. New uses keep cropping up in the journals.

For instance, buckyballs of carbon 60 give film optical properties that are not linearly dependent on the intensity of the light. Instead of slowly getting darker, the film—at a definite point—becomes opaque. Eventually we may have better ways to protect ourselves and our equipment from high intensity light. Efficient regulation of light is a property that has many other practical uses. Fiber-optic networks using all-optical switches and modulators are an exciting possibility. Someday we may have optical digital processors.

Compressed buckyballs make diamonds, but "doped" fullerenes are even more interesting, for they acquire superconductivity when atoms of alkali metals like potassium or rubidium are made to fill the spaces between the carbon molecules. In Japan chemists glue the molecules together with palladium to make a polymer that's unusually stable and electrically neutral.

Gas molecules like oxygen will fill the octahedral spaces between buckyballs that are packed together like crated oranges. It turns out that these packed fullerenes are selective —the gas molecules can't be too large or too small. A useful product would be membranes permeable only to certain gases.

Since fullerenes are like soccer balls, they are hollow and can be called "cages." When scientists aren't coating the outside of the cage with different atoms, they're trying to put something inside the cage. They've used potassium, titanium, iron, scandium—it's astonishing. What good is it to have a different atom trapped inside a big carbon molecule? Depending on what's inside, these fullerenes may make good chemical catalysts or have useful magnetic properties. It's also hoped that medicines coated this way could be delivered into a tumor without hurting the rest of the body on the way. Exploration is under way into the many other aspects of fullerenes used as cages.

There's been a lot of shouting about buckytubes. These are thought to be fullerenes that form into helical-structured tubes instead of balls. People hope they'll make strong fibers. The only trouble is that some scientists have succeeded in taking pictures of fullerenes with scanning tunneling microscopes. The balls are there, but when one group tried to find buckytubes, they couldn't.

In the meantime, the rest of us can sit back and watch with fascination as the results on fullerenes pour in and postulated practical applications become reality.

Super-Diamond

The diamond is so much a supersubstance in so many different ways that it might seem there is no possibility of a super-diamond. Nevertheless, that is evidently what chemical engineer William F. Banholzer, of General Electric in Schenectady, obtained recently.

What makes a diamond a supersubstance? What, for instance, makes it the hardest substance known?

A substance is hard when the atoms that make it up cling tightly together so that it is difficult to jar them loose by impact. Thus, when diamond is used to scratch other hard substances, atoms are jarred loose from the other substance by the diamond, but nothing jars the diamond atoms loose from each other.

To make atoms stick together, each one must hook onto as many neighbors as possible. The tightest packing is where each atom hooks onto four others. The smaller the atom that hooks on in this way, the closer those atoms come to each other, and the more firmly they remain in place. The carbon atom is the smallest atom capable of hooking onto four neighbors, and the diamond is made up of carbon atoms only.

Carbon atoms do not always pack together in the closest possible arrangement, however. They are loosely packed in coal, coke, and soot, for instance. They are also loosely packed in graphite (the "lead" of lead pencils). However, if a piece of graphite, say, is heated to a high temperature so that the atoms can rearrange themselves and is also put under huge pressures so that the atoms are forced as closely together as possible, the result is diamond.

Deep underground, the temperatures and pressures are high enough to produce diamond, and in a few places, such diamonds have worked their way close enough to the surface to be found. It was not until the 1950s that scientists at General Electric were able to devise methods of getting temperatures and pressures high enough (in the presence of certain metals) to get synthetic diamonds that were identical to the natural product.

Diamonds are transparent and bend the light that passes through them into myriads of rainbows. If one is cut into the proper facets, it shows flashes of different colors ("fire") as it turns, even though it is a colorless substance. This is incredibly beautiful, and that is its chief value to people generally.

In industry, however, its hardness makes it the best abrasive we know for rubbing down and smoothing out unevennesses. (There are abrasives that are cheaper and almost as good, but they are only *almost* as good.) Then, too, though diamond does not conduct electricity, it does conduct heat very easily. This means that the use of bits of diamond in microchips or in other miniaturized components of today's devices will keep them from overheating. In addition, of all transparent substances, diamond is least likely to be damaged by radiation, which means that it could be very useful in devices involving laser beams.

It would be nice if you could get something that conducts heat and resists radiation to an even greater extent than diamond does, but since the diamond arrangement of atoms is the ultimate, how can you do better?

Well, there are two kinds (or "isotopes") of carbon atoms: carbon 12 and carbon 13. Every bit of carbon, including all diamonds, contains about 99 percent carbon 12 and 1 percent carbon 13.

Both types of carbon act almost precisely the same chemically, but carbon 13 is about 8 percent more massive than carbon 12, and that does introduce certain tiny differences. About fifty years ago, for instance, a Soviet physicist argued that when heat passes through a diamond, it does so most efficiently when all the atoms are of the same type. Even if all are carbon, the occasional carbon 13 trips up the heat energy somewhat and slows it down.

Scientists have developed ways of separating isotopes of elements, and it is possible to begin with methane, a carbon-rich gas, and to treat it so that virtually all the methane molecules containing carbon 13 are eliminated. You then have methane molecules containing carbon 12 atoms only, and this methane can be used to make synthetic diamonds that are carbon 12 only. This was first done by a team under Banholzer at General Electric in 1988.

Carbon 12 diamonds have now been tested by scientists

at Wayne State University in Detroit and found to be super-diamonds indeed. The carbon 12 diamonds can conduct heat 50 percent better than ordinary diamonds can, and they can withstand ten times the intensity of radiation. What was thought to have been the ultimate has been beaten.

Unfortunately, carbon 12 diamonds are even more expensive than diamonds ordinarily are, so that it is not to be expected that they can be used on a large scale. However, there are bound to be small-scale uses for which they will be ideal, and it is quite possible they will be essential to computers and lasers of the future.

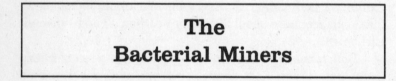

The Bacterial Miners

A group of scientists at the Idaho National Engineering Laboratory has reported on the use of bacteria to help release cobalt from low-grade ores.

This could well represent a new and extremely important technique for carrying on the search for metals—something that human beings have been engaged in for five thousand years. It was metal that brought the Stone Age to an end, for metals could be easily beaten into shape, whereas stone had to be chipped or ground. Metals were harder and tougher than stone tools, less easily breakable, and held a more permanent edge.

There was only one catch. Metals were far harder to find than rocks were. The very word *metal* is from the Greek word meaning "to search for."

At first only those metals could be used that could be found in free form—gold, silver, copper, and meteoric iron. When it was discovered that certain rocky "ores" could be heated and smelted and forced to yield metals, the supply of these desirable materials increased. Metallurgy therefore became one of the marks of early civilization.

Even so, metals remained scarce. Copper itself was too soft to use for tools, but when mixed with tin, it became "bronze," which was much harder. Bronze was used for tools and weapons (the "Bronze Age") until efficient methods for smelting iron ore were discovered.

Tin, however, is quite rare, and the sources of tin ore available to early civilizations were just about used up by 1000 B.C. This was the first case of the actual disappearance of a resource. The Phoenicians traveled out into the Atlantic Ocean and discovered tin ore in the "Tin Isles," which may have been Cornwall in England, and grew rich by monopolizing this valuable material till iron made it comparatively obsolete.

Since then, human beings have searched the world over for natural concentrations of ores containing desirable metals. It was not only a matter of gold, silver, copper, and iron, but newer metals discovered only in modern times. Nickel, cobalt, tungsten, manganese, chromium, vanadium, niobium, and so on—all had their valuable uses.

High concentrations of the necessary ores were dug up and smelted. The metals were then used and, eventually, discarded as scrap. The highest concentrations were used up, and metallurgists learned to use poorer concentrations and get the metal out of them economically. The metals are never "used up," but they are spread out over the Earth's surface in lower and lower concentrations, and getting more of these materials becomes an ever-increasing problem.

Many simple life forms are more able to concentrate certain elements out of very low-grade sources than any human technology can. For instance, iodine is a useful element that is very rare. It occurs in seawater, but in such low concentration that chemists cannot extract it profitably. However, seaweed can do it. The plants filter the rare iodine out of vast quantities of seawater, and then human beings can harvest the seaweed, burn it, and isolate iodine from the ash. Seaweed is our iodine miner.

There is a bacterium, called *Thiobacillus ferroxidans*, that actually lives on minerals. It extracts sulfur atoms from the minerals and combines them with oxygen to obtain the energy it lives on. In so doing, the bacteria break up the mineral and release other types of atoms in them. This can result in the liberation of such poisons as arsenic and cyanide, so that to miners, the bacteria could be dangerous.

However, under properly controlled conditions, particular varieties of this bacterium were found to flourish on ores that contain cobalt in concentrations too low to be economically smelted by human beings. When the bacteria have gnawed away at the mineral for a while, 10 percent of the cobalt content can be easily dissolved out. So far, the people at the Idaho National Engineering Laboratory have made this work only on small-scale samples, but there's no reason it can't be made into a large-scale technique, eventually.

It may be possible to develop strains of bacteria that can live on a wide variety of ores so that metals of all sorts can be obtained from sources that would ordinarily be given up as uneconomic. In fact, there may be more than that. We might, conceivably, set them to work on metal scrap that has been discarded and cannot be profitably recycled.

The industrious little bacteria may live on the scrap if it is mixed with the proper kind of what is to them nutritious rock, and they will then recycle the metals for us. In that case, we will *never* run out of metal.

The
Park Phenomenon

Other creatures select territory best suited to them, and many create homes for themselves. We human beings do this too, but we also select and create places not to live in, but for what we have labeled "recreation." These places are now called parks, and most of us feel they recreate our health and sanity.

The kind of parks we have in the late twentieth century did not always exist. The first known parkland used for recreation belonged to an ancient Persian king who liked to hunt for pleasure. Down through the ages, the estates of most monarchs had gardens and hunting areas, and sometimes the poor public was allowed in. As the common people became more powerful, they used and built parks for themselves. The first public park in the United States was Boston Common, dating from 1634.

Most countries have set aside wilderness areas, and most cities contain plots of land with grass and trees. It's not a funny coincidence that big and little parks often closely resemble the savannah-forest habitat where our ancestors evolved. Primitive hunter-gatherers lived in what we would easily recognize as a parklike environment. Even agriculture fits in. A primitive probably invented it because she thought it would be easier than trudging all over to gather plants that might or might not be available. Eventually humans discovered not just how to plant and reap, but how to work with living plants. They made choices about what grows, not just what to gather. The true garden was born.

In many countries, parks with gardens took on a highly formal appearance, as if humans wanted to prove that they could control nature completely. In England, these artificial savannahs changed character in accordance with ideas of "naturalism." Parks, especially on large estates, were designed to seem "natural," with curves instead of the straight lines favored in Versailles.

There is one outstanding park that exemplifies a deliberately created, naturalistic savannah-forest-garden environment. It's called Central Park, and I am fortunate to live right next to it. Central Park, as New Yorkers are fond of saying, is roughly the size of the kingdom of Monaco. The great landscape architects Olmsted and Vaux designed and built Central Park from a wasteland filled with squatters' shacks and garbage. In Central Park there's only one straight line—the Mall, bordered by trees, ending in the famous steps leading down to Belvedere Fountain and the boating pond.

Central Park has it all—an undulating landscape, punctuated by trees and rock outcroppings, streams and ponds, various lawns and fields. Any hunter-gatherer would be pleased, although nowadays the only hunters are unfortunately humans who prey on other humans. There are, however, plenty of gatherers from various ethnic groups harvesting such delicacies as wild garlic, lamb's-quarter (that's a plant), knotweed, june and hawthorn berries, cherries and apples and nuts (which one must share with the birds, squirrels, rats, woodchucks, and perhaps a raccoon).

With the exception of the fantastic and climbable Alice in Wonderland statue, the most popular statuary in Central Park is not human—the ugly duckling next to Hans Christian Andersen; Kemeys's mountain lion crouching above the east drive; Balto, the Alaskan husky who led the dog team to deliver diphtheria serum "to the relief of stricken Nome" in 1925. And of course there's a zoo, now newly revised and more naturalistic than ever, especially when you walk through the tropic house and the birds whiz by your ear and peer at you from tree limbs within reach.

Our primitive ancestors had no idea of the marvelous history of our planet, a history often beneath their feet and readable if you know how. In Central Park, the outcroppings of Manhattan schist—a metamorphic rock formed around 500 million years ago (and so tough it's suitable as the bedrock for skyscrapers)—contain grooves left behind by the ice age glacier.

Some say that Central Park is overwooded, the trees not thinned out as they were meant to be. Perhaps many of us like lots of trees because our ancestors were primitive hunters in forests, while peoples in less wooded lands were building Babylon or arguing philosophy in the Agora.

To the homesick hominid inside a city dweller, a park—especially Central Park—is a place of respite with scenery to rest the eyes and mind, its air seeming less polluted, richer in oxygen. Olmsted planted trees around the park's perimeter to screen out the buildings beyond. In some areas of the park, notably the hilly and forested "Ramble," it's hard to believe you are in a huge city until you hear the distant hum of traffic. It sounds like a steel-and-concrete animal curled around the park, breathing heavily.

Saving the Species

We continue to hear news about efforts to find living examples of species that might be extinct and to save those that are in danger of extinction. This desire to preserve the diversity of life is a pleasant comment on the civilized instincts of at least some human beings.

It was not always so. The people of the Stone Age are at least partly responsible for the disappearance of the magnificent mammoths and other large mammals of the time. In modern times we have seen the careless or even wanton destruction of fascinating life forms by human beings or by pet cats and dogs. The dodo, a large flightless pigeon on the island of Mauritius, was killed off, as was the great auk of the northern seas—the Arctic version of the penguin. The Stellar sea cow, largest of all sirenians, was casually slaughtered, as were the passenger pigeons of North America, who once filled the air with flights of billions. The huge bison herds of the West were ruthlessly killed, partly in order to defeat the Indian tribes who depended on them for sustenance, and were driven nearly to extinction before action was taken to save the survivors.

Times have changed. People work tirelessly to save such endangered species as the whooping crane and the California condor. Sometimes such species can be saved only by capturing them and sequestering them in a zoo where they will be free of predation. Headlines are made whenever a condor egg is hatched in a zoo, for instance.

In Tasmania, the island off the southeast coast of Australia, there was once a carnivorous doglike mammal that is not related to dogs at all. It was a pouched mammal, a marsupial, related more closely to the kangaroo or the koala. Its hindquarters were striped, and it was called the "Tasmanian tiger" for that reason, though its more scientific name is *thylacine*. It was wiped out after Europeans came to Tasmania because it preyed on the sheep. The last-known *thylacine* died in a zoo in Hobart, Tasmania's capital, in 1936. No living *thylacine* has been definitely seen since, though there are occasional reports of fugitive glimpses of these creatures in the wilder parts of the island, and scientists try to track down the elusive animal by computers programmed to find the spots most congenial to the known requirements of the *thylacine*. If any are found, every effort will be made to declare the region off-limits to people, and, if possible, some specimens, male and female, will be taken to zoos where it might be possible for them to breed. (Many animals do not readily breed in captivity. Apparently, some stimuli are required that act only in the wild.)

In the forests of Madagascar lived the largest of all birds—the *aepyornis* (or "elephant bird"), which weighed up to a ton, laid the largest known egg, served as the model for the "roc" of the *Arabian Nights,* and was driven to extinction even before Europeans arrived. Its legend eclipsed a much smaller Madagascar animal also thought to be extinct, but recently found by Bernhard Meier of Ruhr University. He found the "hairy-eared dwarf lemur," an animal that had not been seen since a skin was turned over to scientists in 1964.

A lemur is a primitive primate, the class of animals to which human beings belong. This dwarf lemur is the smallest known primate, with a body (exclusive of tail) only five inches long, weighing about three and a half ounces. It is probably the closest living example of the kinds of primates that existed when the dinosaurs vanished. Undoubtedly, every effort will be made to preserve its existence.

On New Zealand, there once lived the giant moa, the tallest bird that ever lived, though it was lighter than the *aepyornis.* Some moas were twelve feet tall, but they were driven to extinction by the Maoris before the Europeans arrived.

In New Zealand there also lives the kakapo, a large, flightless, nocturnal parrot, once widespread in the islands but now driven nearly to extinction. The bird only breeds once in four years and only when the food supply is good, so it is hard to keep it going. Only forty-three birds are left, fourteen of them female. Some years ago, twenty-two kakapos were carried off to a small offshore island where there were no cats or other predators.

There extra food was brought in so that the birds, growing fat and comfortable, might breed. When one of the kakapos laid an egg, *that* made the headlines.

Plant Power

We've come almost full circle with tomatoes. South American Indians ate them, but when the Spaniards brought the plants back to the Old World, Europeans called tomatoes poisonous "love apples." It took centuries for the tomato to become the popular food item it is today. Or was, for according to news headlines, tomatoes are now suspect. Some of them have been (sharp intake of breath, a shiver) *bioengineered!* What it means is that tomatoes stay fresh longer because an installed "Flavr Savr" (!) gene prevents the tomato's built-in softening gene from going to work.

Other biotech products are coming. Because some inserted gene material may prove to be allergenic, everything should be tested, but hysteria is pointless, especially since supermarket fruits and vegetables are already doused with pesticides and coated with wax.

Since agriculture began, humans have been bioengineering foods, selecting seeds from the best plants, producing bigger and better fruits and vegetables. In fact, one of humanity's problems is that we've overdone the selection process and are losing plant diversity, which is vital. As an environment changes, different plants may prove more adaptable than the ones always grown.

Various seedbanks have been established to preserve the invaluable diversity, but almost all of them contain too few varieties and are in trouble because too many countries still believe it is more important to fund the military than to protect future generations. Now the Norwegians have started the Svalbard International Seedbank deep in an old coal mine on Spitsbergen, an Arctic island, which will keep the precious seeds frozen—and ready in case other parts of Earth experience plant extinctions.

Prehistoric humans made good use of the wide diversity of plants. While they ate a great deal of meat, it was tough and lean, with little saturated fat. The fish and birds they caught and the nuts, fruits, and vegetables they gathered made for a remarkably healthy diet. Jane Goodall has shown that chimpanzees are also omnivores, but the chimpanzee, like the gorilla, has a much bigger large intestine than a human. We have a harder time digesting a high-fiber totally vegetarian diet, which doesn't provide enough protein if you carefully wash off all those nutritious insects infesting your dietary plants.

Plants are such a valuable source of medicines that it is even more imperative to maintain the diversity of wild plants. The only way that can be done is to educate humans to stop environmental destruction. Earth's forests are threatened by the engulfing tide of greedy human overpopulation, and those forests contain not only myriad sources of medicine already in use, but many plants only now being investigated. In Belize, botanists Michael Balick and Robert Mendelsohn found that land value is eventually destroyed when forests are cleared for poorly productive ordinary agriculture, but goes up when the land is used for medicinal harvesting. In Guyana, a UN project to save the rain forests is now under way, with prospects of finding valuable plant genes for bioengineering as well as useful pharmaceuticals.

New uses are being discovered for ordinary, nonbioengineered plants. A variety of pawpaw growing in the eastern United States contains an anticancer drug called "bullatacin," now under investigation. In India and Burma, leaves from the neem shade tree have always been used to kill insects, so modern researchers are studying the neem's pesticidal compounds (called "limonoids"). In South Africa, the "fynbos" type of scrubland contains a huge diversity of plants (1,470 different ones on Table Mountain!) with many medicinal uses, but all are threatened by imported vegetation. There is now much controversy going on over methods to control the imported vegetation, for settlers in South Africa have come to depend on the new plants and trees in many ways. Planetary ecology is a highly complicated business.

Humans are not the only ones who use plants for more than food. Some birds line their nests with leaves that help control parasitic infestations. Bears chew up an herb to rub it on their fur, presumably to control insects. The main animals to use plants consciously and deliberately for medicine are chimpanzees, our closest relatives. Chimps are highly selective about eating certain plants when they are feeling under the weather, and the plant selected usually fits the medical problem. Then there's that one elephant who was observed to eat tree leaves that Kenya natives brew as a medicine to induce labor. The elephant gave birth, leaving researchers wondering whether she ate the leaves as medicine or by accident.

While we study the uses other animals make of plant power, and try to preserve the incredible variety of plants we might need someday, it is still important for us to experiment with our own technological ability to change, improve, and protect plants. So when you hear about a scientist zapping a plant with a "particle gun," don't be alarmed. Perhaps the scientist is only trying to help the plant resist disease so it doesn't have to be grown with so much pesticide. You may end up with a better, healthier, longer-lasting, and less-polluted fruit or vegetable.

Plant Help

Everyone would suffer if all plant life suddenly disappeared. Creatures who eat plants would die first, and then the carnivores who eat the vegetarians. Down in the ocean depths there would still be animals living on the bacteria feeding on chemicals from volcanic vents, but the land would be a bacteria-laden graveyard. Humanity, if wise, would have stored seeds in a space settlement; if not, a few desperate human beings left on Earth might try growing vats of bacteria for—ugh— food while trying to hurry up artificial evolution to develop plants again.

Human beings are omnivores and always have been, despite notions about ancestors that lived by hunting. The fact is, prehistoric *Homo sapiens sapiens* was a gatherer of plant food as well as a hunter. So was *Homo sapiens neanderthalensis*, who in addition to banqueting on forest fruits and nuts, put flowers on the funeral biers of loved ones.

Lately, the botanical side of nature has been prominent in the news. Many types of cancer are statistically less frequent in people whose diets are full of vegetables, fruits, and nuts. It's said that chemicals in cruciferous vegetables like cabbage and broccoli lower the risk of breast cancer by increasing the metabolism of estrogen, breaking it down to an inactive form.

As rain forests dwindle, botanists search them for marvelous plants to help humanity. Farther north, a conifer called *Taxus*, or yew tree, is achieving notoriety. Its bark contains taxol, useful as an anticancer drug because it stops cell division.

On a less medicinal level, scientists like Sharon Long at Stanford study plant growth with an eye to what will help humans. Long's laboratory has produced a test for the most efficient varieties of alfalfa and is investigating how alfalfa works out its symbiotic relationship with the bacterium *Rhizobium meliloti*, which comes to live in the root nodules. The botanical know-how from these studies may help to engineer fertilizers and pesticides that would be produced only at the roots of specific plants, without effect on other plants.

Plants help humans in all sorts of ways, but what's being done to help plants? Fighting to save the dwindling green areas of Earth is the chief battle, but there are other, smaller ways of helping. For instance, the greenhouse production of tomatoes for British markets has been hampered by the fact that honeybees don't like tomato flowers much and are picky about temperature and sunlight, neither of which is reliable in northern countries. Artificial pollination by human workers is costly, inefficient, and often damaging to the tomato flowers. It's been found that a homely, less social bee called the "earth bee" works longer hours, longer months, and with no prejudice against tomato flowers.

The plant world would be much healthier if human beings didn't pollute the air by burning nonbiodegradable plastics. Recently, American genetic engineers have inserted bacterial genes into thale cress, which then makes polyhydroxybutyrate, a biodegradable plastic. If someday we can grow "good" plastics, we could stop manufacturing plastics that stay forever in landfills, kill our sea life, and pollute our air. It would also please me to think of farmers plowing under their tobacco farms to grow plastic instead.

According to an Edinburgh group of scientists, farmers may eventually be able to find out early that their crops are in distress. In each field there would be a few "biosensor" plants to signal danger and the need for help. Already produced are genetically engineered plants that contain a protein called "aequorin," originally found in a jellyfish. The scientists put the aequorin-making gene from the jellyfish into a bit of DNA, which is then inserted into what they call a "Trojan Horse" microbe. The microbe (*Agrobacterium tumefaciens*) parasitizes the plant, adding the gene to the plant's genes. The result is a plant that emits a blue light under stress.

Efforts to save trees are making progress. Molecular biologists may help bring back the magnificent chestnut trees that provided shade, beauty, and food for early colonists in America. The devastating chestnut blight (a fungus) can be stymied by infecting it with the gene of a virus that renders the fungus less virulent.

Humans waste paper, which represents vast quantities of demolished trees. After use, paper can be recycled. Now the manufacture of paper can be improved. Wood pulp is ordinarily processed by removing its carbohydrates and then flattening and drying it. Chemical engineer G. Graham Allan at the University of Washington says that the resulting waste space in wood pulp could be filled by soaking the pulp in solutions that would precipitate calcium carbonate into the spaces. In this method, 30 percent of the pulp could be replaced by filler, thus saving enormously on the number of trees needed to create the pulp in the first place.

If my emphasis on helping plants seems to be rebounding to aid for humanity, that's the point—we're connected. We're all *one life*, on *one planet*!

Underneath

Japan, overcrowded, with land values climbing into the stratosphere, is thinking of moving underneath—below the ground. The Japanese are planning to begin with underground sewage plants, then underground railroads, finally underground cities.

It's not so unthinkable, actually. We have underground railroads called "subways." Cities like New York have a sub-terranean world of electrical wiring, sewer lines, gas mains, and so on. In northern cities with long, harsh winters, there is a tendency to build underground shopping malls, veritable cities in themselves.

The thought of living underground, of burrowing in the Earth like moles, of separating oneself from the air and sky may seem unpalatable, but, if we stop to think of it, there can be many advantages to living underground.

First, weather would no longer be important, since it is primarily a phenomenon of the atmosphere. Rain, snow, sleet, fog would not trouble the underground world. Even temper-ature variations are limited to the open surface and would not exist underground. Whether day or night, summer or winter, subtropical or subpolar, temperatures underground would be in the neighborhood of 55 to 60°F. The vast amounts of energy now expended in warming our surface surroundings when they are too cold, or cooling them when too warm, could be saved. The damage done by weather to humanity and its structures would be gone. Even earthquakes would be only about a fifth as damaging beneath as on the surface.

Second, local time would no longer be important. The passage of the sun, the tyranny of day and night coming at different times in different places can be avoided. Underground, where there is no externally produced day, the alternation of work, play, and sleep can be adjusted to suit ourselves. The whole world could be on eight-hour shifts, starting and ending on the same stroke everywhere, at least as far as business and community endeavors are concerned. This could be important in a freely mobile world. Air transportation over long distances, east and west, would no longer entail jet lag. If we leave New York at noon and take twelve hours to get to Tokyo, it will be midnight there, and midnight for our own biological clock, too.

Third, the ecological balance of the Earth would be better

off. To a certain extent, humankind encumbers the earth. It is not only our enormous numbers that take up room. It is also the structures we build to house ourselves and our machines, to make possible our transportation, communication, and recreation. All these things distort the wild, depriving many species of plants and animals of their natural habitat—and sometimes, involuntarily, favoring a few, such as rats and roaches. The more of ourselves and our works we place underground, below the realm of burrowing life, the more room there will be on the planet for other forms of life.

Fourth, nature would be *closer*. It might *seem* that to withdraw underground is to withdraw from the natural world, but would that be so? Would the withdrawal be so much more complete than it is now, when so many people work in city buildings that are often windowless and artificially conditioned? Even when there are windows, what is the prospect one views in some places but other buildings?

One might argue that there is a psychological difference. However divorced present-day cities may be from nature, we are in view of the sun and the sky, by looking out the window or by stepping out the door. Isn't that right?

But look at it this way. To get away from the city now, to reach real tracts of greenery and of reasonably unspoiled nature, a New Yorker, a Londoner, a Tokyoite must travel horizontally for miles and hours, through heavy traffic—first across city pavements, and then across suburban sprawls.

If we were living beneath, if we had an underworld culture, the countryside would be right there, a few hundred yards above the upper level of the cities—wherever you are within them. The world of nature would be an elevator ride away, and the dwellers beneath would see more greenery, under ecologically healthier conditions, than dwellers of surface cities do today.

And remember, as one more point, that in an underground world of perpetually equable weather, walking would be much more pleasant. There would be less reason to use transportation facilities for short hauls, thus conserving energy and increasing bodily fitness.

Are there disadvantages to life underground? A few. There would have to be a huge capital investment, a great psychological adjustment. There would be the problem of the vast ventilation procedures that would be necessary and, of course, the danger of fire—which may do more damage in the caverns than in the open.

By the way, once again I am not a totally dispassionate observer. I happen to like enclosed spaces, and back in 1953, I wrote a novel entitled *The Caves of Steel* in which I dealt with and described an Earth made up entirely of underground cities.

Business Leads the Way

"Smart cards" are becoming more and more important. These are, essentially, computerized objects that can be made small enough to look like credit cards, or can come in many other handy shapes. They can contain enough information about you, personally, to identify you absolutely—your fingerprints, your retinal pattern, your voiceprint, and so on—so that they come close to being absolutely secure, usable by only one individual. They can also contain enough economic information to make it possible to carry through business transactions with great ease, transferring funds, keeping track of the state of one's assets, and so on.

This is the latest development of a long series of changes

that have worked to make it easier to trade and, therefore, to do business.

At the start, there was barter, in which two individuals bargained some sort of exchange that would benefit both. This was slow, and there was never any certainty that, in the end, one, or both, would not feel cheated.

A medium of exchange was developed in the form of rare metals, such as gold or silver, and material goods could be valued in them. Wealth became portable. Coins were invented in Lydia in the eighth century B.C., the government producing gold and silver in weighed lumps of guaranteed quality, stamped to make it official, so that there would be less opportunity to cheat.

Paper money was invented in medieval China so that wealth could become still more portable. In late medieval Europe, banking houses were invented, and notes from one to another made financial transactions nationwide. Checks were invented, which were individual notes, pieces of paper representing any quantity of money. Credit cards made it necessary to make out only one check a month.

And now smart cards.

One could make the argument that the need to facilitate business was the driving force behind technological advance. The Phoenicians invented the alphabet because they were a trading nation dealing with the great Babylonian civilization to the east and the great Egyptian civilization to the southwest, each with a terribly complicated written language. The alphabet was a form of shorthand into which either or both languages could be translated and trade facilitated.

The development of roads and canals and ships was not so much for people in general to travel (people didn't, until contemporary times) as it was to make it easier for goods to be transported. The same might be said of more recent inventions in the field of transportation and communication, from steamships and railroads to airplanes and radio. Every advance increased the volume and scope of trade.

This is important, of course, because no individual, no small group has everything or can make everything. As a result of trade, food, manufactured objects, and knowledge can be exchanged, and each individual or small group can find an opening to a wider world, both physically and mentally.

We have now reached the stage where the world is a single economic unit and where every group has access (some more than others, of course) to the products of all other groups.

There is no way of overestimating the importance of this. We live in a time when the problems and dangers humanity faces are worldwide. The dangers of nuclear war, of chemical and radiational pollution, of overpopulation, of desertification, of the vanishing ozone layer, of the greenhouse effect are *global*. No nation can escape.

Efforts to solve these problems must also be global. No nation by itself can effectively deal with any one of them if other nations do not cooperate. There are continuing international meetings on all such problems, for that reason.

How can we persuade nations, divided by thousands of years of national traditions that dictate rivalries, suspicions, hatreds, and wars, to cooperate? We must find some common interest. Language, religion, culture in general are all divisive. Science is a unifying factor, but few people feel an overwhelming interest in science.

That leaves business. These days, anything that interrupts free international trade harms everyone. Most of the world is aware of the danger of protectionism, and objections arise to forms of protectionism that still exist in Japan and elsewhere.

Already large business firms are multinational and are forced to think in global terms. Narrow considerations of nationalism and patriotism simply don't make sense. This must continue if we are to survive, and anything that facilitates trade and continues to make business more international necessitates a further increase in global thinking.

This must make it easier to solve the world's problems, if

they are to be solved at all. Business, like it or not, leads the way.

Checkmate?

Not too long ago a world champion of chess, Gary Kasparov, easily defeated a computer in two chess games. Many people breathed a sigh of relief. A human being had beaten a computer, and human superiority still existed.

Actually, this reaction is wrong in several ways. First, the victory is only temporary and is not likely to be maintained. Chess-playing computers are not very old—just a few decades. They have become better and better as the programming has become more intricate and efficient and as the computer capacity for taking into account myriads of possible chess moves has been steadily expanded.

Right now the chess-playing computer has beaten other human grandmasters, and it took a world champion to defeat it. The two losses will be carefully studied by the computer experts, and the weaknesses in the programming revealed by those losses will be corrected. The next time out, Kasparov will have a harder time and might well be defeated.

But what if that happens? It means nothing from the standpoint of "superiority." Already computers can perform mathematical feats in a short time that human mathematicians would require lifetimes to duplicate. (Even cheap pocket calculators can solve arithmetical problems faster than human

231

beings can.) Computers can not only work millions of times faster than ordinary mathematicians, but can do so without likelihood of error.

That, however, doesn't mean that computers are "smarter" than mathematicians. They can only manipulate numbers on command more rapidly, more accurately, and more tirelessly. The mathematician must still command the details of the manipulation.

Nor does the ability to play chess represent an essential advance beyond this. Chess is a severely limited game. It is played on a board of sixty-four squares with thirty-two pieces of six different types, and each piece can move only according to a fixed and limited pattern. To be sure, there are enormous numbers of individual chess games possible, but a man like Kasparov studies chess constantly and has memorized large numbers of openings, closings, and midgame situations, so that in some respects (not all) he plays mechanically. A computer can, in principle, do this with greater mass memory, and therefore it can, eventually, outmatch any human being. If it does so, it no more shows any real superiority than when it solves vast numbers of differential equations simultaneously.

In fact, a computer victory would not even wipe out chess as a competitive sport. It would merely mean that there would eventually be matches between human beings and other matches between computers. We see this happen in the field of racing. A man on a horse can run faster than any human being, so naturally there are horseraces in which human beings, running on foot, cannot compete. But there are also footraces, just as eagerly contested. And, for that matter, there are automobile races in which horses cannot compete.

But might it be that chess playing is just a symptom and that, eventually, computers might outdo human beings in any form of intellectual achievement? I don't think this will happen.

The human brain contains 10 billion nerve cells and 90 billion supporting cells. It will be a long time before computers ever have that many units. Nor is it just a matter of units; each cell in the human brain is connected with unimaginable intricacy to a large number of others in a pattern we don't understand. Finally, the cells are not mere on-off switches as the computer units are. Each cell contains millions of large and complex molecules with whose intimate functioning we are still unfamiliar.

It will be a long time before a computer can imitate the complexity of the human brain, and there is not much sense in striving to reach that goal. It would be easier to improve the human brain itself through the techniques of genetic engineering and to confine computers to do their ever more efficient task of number-crunching.

But will our complex human brain have anything to do after we leave number-crunching to the computer? Of course it will. The games of art, of literature, of scientific research, and many other things of that sort are, as far as we know, limitless. The individual pieces are enormous in number, the combinations beyond computation. Human beings can perform as artists, writers, scientists, musicians, inventors, and so on and so on, not by any easily described procedures, but by the use of unknown processes we call "intuition," "insight," "imagination," "fantasy," and the like.

We cannot describe such processes accurately in terms that will allow us to program a computer to duplicate them, because we don't know how our own brains manage it.

Given that, then, human beings will always have a place that the computer cannot reach. We will not be checkmated.

Cockroaches
and Computers

When the first electronic calculators appeared—and wiped out sales of Isaac's then-new book on the slide rule—he said in print that for all he knew, the things worked because each contained a particularly lively and intelligent cockroach.

Thirty years later the description doesn't seem far off. The original remark was prompted by fond memories of Don Marquis's famous Archy, but that literary insect was intelligent because he was a poet whose soul had unfortunately "transmigrated" into the body of a cockroach. Real cockroaches are not poetic.

Are there any similarities between cockroaches and computers? To begin with, cockroaches have been around more than 250 million years longer than humans have existed, and computers are younger than many of us. The cockroach pedigree is impressive—phylum Arthropoda (on Earth for 630 million years), subphylum Uniramia, subclass Ptergota, class Insecta. Like all arthropods, cockroaches have a hard outer skeleton, and like terrestrial insects, they breathe through air ducts containing valves. Their nervous system is a double chain of nerve cells in bunches called "ganglia." The biggest ganglion in the head receives sensory input and sends instructions on, but the ganglia in the thorax and abdomen are also large enough to contract muscles even when the head has been thoroughly stepped on.

Exoskeletons and air ducts insure that all insects have to be relatively small. Large brains can't develop in very small animals, but in the aggregate—as if the individuals were linked like chips in a computer—some insects do astonishing things.

Social insects like bees, wasps, and ants build complex societies with intricate communication. Cockroaches don't, but scientists believe that they are the ancestors of an impressively social insect—the termite, adept at building large edifices of their own, or tearing ours down.

Cockroaches seem like little machines, but they learn faster than our machines do. Sensitive to vibrations of air and surfaces, they avoid us, and even learn to avoid places where you've usually put poison. An individual cockroach lives over four years and may well lay a thousand eggs. Cockroaches are found almost any place where humans can live. However unintelligent, cockroaches are successful. So are computers.

I use a computer but agree with Isaac that there must be an Archy inside. I called my brother, a computer maven, and learned that modern computers are based on integrated circuit silicon chips that look like little squares with a lot of tiny legs (more like millipedes than cockroaches, which make do with six legs). They don't run under the refrigerator because the legs (for input and output) are soldered into a circuit board that contains thin metal strips to connect the chips.

A chip is made of thin silicon layers treated to give the right electronic properties. Circuit microphotographs are used to etch circuits on the chip's layers, so many thousands of electronic circuits go on one chip, enabling it to handle a great deal of input and output. A board may contain many chips. With new techniques using beams of electrons and X rays to etch circuits, components will be made smaller so even more will be packed in, increasing the computer's power.

My brain went into overload by the time my brother got into logic gates (out of four types of logic gates—like valves —you can build any computer). I kept picturing cockroaches as simple but efficient little machines, and their termite descendants as chemically wired-together chips.

Lately I've been reading about people making "silicon neurons," analog integrated circuits that approach the functional characteristics of genuine nerve cells, which handle ion currents to produce nerve impulses. Some scientists are even now trying to combine many silicon neurons onto one microchip, connecting the chips to imitate the way organic brains process information.

It seems that we're on the way to making what Marvin Minsky has called a "neural net." Computers usually work on a problem linearly, one thing at a time, but a computer with a neural net can assign parts of a problem to little processors that are connected to all the others.

If communication organizes the activities of the neurons in a human brain or the individuals in a termite hill, the ease of communication in an artificial neural net should do wonders for computers. They'll learn faster, and better. Soon these artificial gizmos will imitate brains—but whose?

The microchip is theoretically immortal, but it can be damaged by heat or cracked if sufficiently assaulted. One cockroach can be stepped on, many poisoned, but the clan is all too likely to outlast the human one. In science fiction, intelligent computers outlast humans too.

Cheer up. Unlike cockroaches, microchips don't know how to reproduce by themselves. Yet.

Robettes

MIT, thanks to the presence of its artificial-intelligence guru Marvin Minsky, is one of the important centers of advance in robotics. MIT's Australian-born professor Rodney A. Brooks is approaching robotics from a new angle—the small.

The kind of robots that have been pictured in science fiction stories, both in the print and in the visual media, have always rather resembled human beings. They have been humanoids, artificial human beings. That was indeed the whole rationale for robots in the days before miniaturized computers were invented—to have mechanical human beings who could do the work of the world.

Yet it is easy to argue that human beings are not a proper model for robots. Human beings are all-purpose mechanisms, with generalized limbs and a highly specialized brain. A human being is designed to do many things, from pushing a loaded wheelbarrow to composing a symphony, from chopping down a tree to alphabetizing a set of file cards—and doing it all with the same brain and the same set of muscles. This is, indeed, a terrible drawback, since the exigencies of life usually compel a person to spend most of his time and his magnificently generalized body in becoming specialized. Often he is compelled to do jobs that totally underuse his brain and atrophy it. Or he is compelled, by desire or necessity, to use that brain for one purpose and allow his muscles to atrophy and his body to grow soft. Almost always, he learns to do a few things very well and remains a virtual idiot in other respects.

But if we want a mechanical device to do a particular job, why build a generalized structure that would make it capable of doing a wide variety of jobs it will not necessarily ever be asked to do? The attempt to build a generalized structure involves so many complexities and difficulties, the reality may be very far off. On the other hand, a specialized structure for the performance of a special job may be a comparatively simple task to carry through.

Thus, what we now call "industrial robots" are computerized machines. The microchip, developed in the mid-1970s, made it possible to build cheap, reliable, specialized industrial robots that look like nothing at all human but that are essentially computerized arms, or levers, capable of going through a limited variety of motions that will enable them to perform specific tasks over and over again. They do this more uniformly and efficiently than a human arm run by an overspecialized brain can possibly do, and the robots, moreover, never get tired or bored.

Naturally, such robots are going to be made more complex with time, capable of varying their work usefully. The drive for greater generality will be irresistible. In fact, one can imagine at least two reasons why there *should* be an effort to make robots humanoid.

First, there is a vast technology in place already that is designed to fit the human body. Machines can be run in particular ways because the human body can reach and bend in certain ways, because arms and legs and fingers are of certain sizes, and so on. If robots can be devised that are humanoid in structure, they can use the technology that is already in existence. We won't have to have two of them—one for the human, and one for the robot.

Second, a humanoid robot is more apt to be looked upon and felt to be a friend, as well as a coworker. This may be a strong emotional reason for the development of such things.

But Brooks of MIT feels there should also (and preferably) be development in other directions. He feels that a more appropriate model for robot development might be the insect body, which is built altogether differently from our own. Insects are economical organisms with their functioning built into tiny volumes, and, in their millions of species, they have millions of specializations that suit them to one form of life or another. Why not make a flood of tiny robots (which I think of as "robettes") designed to do specialized jobs à la insects?

Brooks himself suggests that small robots may be given jobs such as removing barnacles from the hulls of ships. He sees them conducting explorations on the soil of Mars. He even imagines very small robots that can be injected into a bloodstream to do surgery from within.

I must add a personal note, however. I have been writing science fiction stories about robots for fifty-one years, and it is not easy for any scientist to come up with an idea about robots that I haven't handled at one time or another.

In a story I wrote in 1974 entitled "That Thou Art Mindful of Him," I discussed the possibility of tiny robots. I imagined a small birdlike robot that flew about speedily, with its task that of disposing of insects. Then, too, in a story I wrote in 1988 entitled "Too Bad," I described a miniaturized robot inside a bloodstream that killed cancer cells without harming normal ones. Not bad ideas!

Once and Future Robots

In 1920, history was made by a play called *R.U.R.*, by Czech playwright Karel Capek. Since then, the Czech word *robot*, meaning "serf," has been used to describe a manufactured device able to do the work of human beings. In fiction, a robot usually looks humanoid, but the name has come to apply to many devices that perform the actions of some human parts, like an automated factory's roboticized arms for manipulating tools.

Most fiction, and even most scientific speculation, deals with the future of robotics, a term Isaac first used without realizing that he was coining something new. But Isaac also contributed heavily to the present development of robotics by writing "I, Robot," stories that influenced the career choice of many leading pioneers in robotics and artificial intelligence. I think Isaac would have liked me to bring readers up to date on some of the current developments in practical robotics.

To be genuinely practical, in our terms, a robot should be able to help us in our world. We already have robot devices capable of moving around in outer space, taking pictures for us, digging up pieces of the Moon or Mars. Other robots go down into ocean depths to examine strange creatures living on bacteria that eat the sulfur coming out of cracks in the Earth's crust.

We can't do any of these things ourselves. But whether a robot is venturing into realms denied to humans, or merely helping us out in daily life, it's easier for us if the robot can see the way we see and manipulate objects the way we do. Preferably, a robot should do these things better.

Better is the important word. In Operation Desert Storm, robots on night surveillance showed soldiers things the human eye can't see in the dark—but the images were understandable, as if the human had suddenly acquired marvelous night vision.

Right now Carnegie Mellon's Field Robotics Center is building a robot that can go out onto the space shuttles, inspect the tiles and perform maintenance. Closer to home is the wool-shearing robot project at the University of Western Australia. Sheep can now be sheared in seventeen minutes, which is presumably better for the sheep as well as the wool industry.

Michael Ali is a graduate student at the New York State Center for Advanced Technology in Automation and Robotics (try that in a school song). He's devised a robotic hand so similar in shape and activity to ours that it's easy to learn to use. Operated at a distance, the robot hand mimics what human hands can do, but in places where humans can't go safely. Perhaps someday Ali's robotic hand will be part of humanoid robots . . . but there I go, dreaming about the future.

Back in the present, useful repair robots are being devised. Roboticists at Northwestern University have one that quickly repairs potholes, the bane of car-happy Americans. At the University of California at Davis, a repair robot can both recognize and repair highway cracks.

Improvements in robot vision are under way. Electronic

vision systems now in use are slow, for the video camera's image must be digitalized and analyzed by computer. We don't know exactly how the human eye and brain recognize objects, but we humans are certainly fast at it, often remarkably accurate with a minimum of clues. You can see only a small part of someone's head to be able to recognize who it is. There's a test for human "structural visualization" that consists of presenting a drawing that resembles a put-together jigsaw puzzle with only a few pieces. Surrounding it are similar puzzles with the pieces turned around and separate. The idea is to pick the drawing of scattered pieces that exactly represents the original put-together puzzle. Artists and structural engineers, and even many doctors, are good at this. Machines are not.

Soon robotic devices may be equipped with an optical, not an electronic visual, system. The optics are complex, using laser beams to carry images of the test object and the object with which the test object is to be compared. This "joint transform correlator" can pick out changing objects with great speed. If the developers can only make the thing pass the structural visualization test . . . but so far, it doesn't work well when the objects are different sizes or at different angles.

Joseph F. Engelberger's Transition Research Corporation specializes in robots useful to ordinary human beings. While they are also busy improving robot vision, using "log-polar" electro-optical imaging, their HelpMate delivery robot trundles along hospital corridors, into elevators, supply rooms, and kitchens delivering meals, medicine, and sterile supplies, day or night. Nurses don't have to leave patients to get medicine from the hospital pharmacy, for HelpMate will pick it up and deliver it. HelpMate learns its route and avoids obstacles in its way, including people. When its way is blocked it announces this and asks to have the obstacle removed.

When they teach HelpMate to cook, I'm buying one.

Music, Always

Now that we are members of the, um, mature generation, we find that we give bigger tips when allowed to take taxi rides in silence, or to the accompaniment of music that, like Cordelia's voice, is ever soft, gentle, and low. Unfortunately, most of the time we're inundated with loud, raucous sounds that some people call "music." We firmly believe that this raucous music even stunts the growth of plants, but since we can't find the relevant articles, this may be wishful thinking.

Music in or out of taxicabs is, however, important to human life. The *Random House Dictionary*'s definition of *music* is a good one: "An art of sound in time which expresses ideas and emotions in significant forms through the elements of rhythm, melody, harmony, and color."

The first human music was probably vocal, and began earlier than humans—primates singing in the treetops; hominids chanting across the grassland. *Homo sapiens sapiens* (that's us) made music while painting in those prehistoric caves that scientists have shown to be as resonant as some concert halls. The caves that are not particularly resonant have fewer paintings. The bone flutes and the mammoth bone drums of Cro-Magnon humans have been found, so they were certainly a musical people.

Other animals, of course, do make music of all sorts, from the lovely songs of birds and whales to the vibrations of grasshopper legs. Honeybees don't even have to watch the dance of a bee who's trying to show the others where the honey is. They can listen to the sound and still know.

Mammals are exposed to music of a sort from conception, for the fetus hears and feels the rhythm of the mother's heartbeat. Small puppies can relax and sleep when a ticking clock is placed in their bed. Music therapy is also used for humans. The temporal organization of music helps certain neurological patients organize their own sequential motion, as in walking and even talking. Some patients who can't talk anymore can learn to sing their messages.

In addition to the pleasant release of endorphins, many cerebral functions come into play when the human brain processes music. It's thought that this helps keep mental functioning alive and well (but not if the music is played so loud that the listener suffers hearing loss).

The average human pulse rate is around seventy beats per minute. This is also the average tempo of most Western music. In fact, it's said that the slow parts of Baroque music induce mental and emotional integration (tell that to taxi drivers). Concentrating on the rhythm of music (usually by tapping one's foot along with it) affects the rate of breathing, making it more regular, and faster or slower depending on the piece.

There are interesting developments involving music. People are making "bioelectric" music, recording the sounds made by the electric impulses that result from brain activity or muscle movement. The sounds are processed through a computerized gadget for the use of composers, musicians, and even handicapped people who can be taught this means of self-expression. It's possible that by studying the "music" made by cells, microscopic creatures, and plants, we will understand biology much better.

Then there are those fantastic machines that are not just music synthesizers but also high-powered computers. One of these days one of them is going to start making music on its own.

To say that music is "sound in time" means, for us humans, sound is from the vibration of air. The oceans are not silent, but full of sound, transmitted through the vibration of water. Space is probably silent because there's no air or water to transmit conventional soundwaves, but who knows? Do the fields of the universe vibrate?

In one novel (Janet's), musicians program a computer with the music made by microorganisms on an alien planet. The computer (so intelligent it could be one of Marvin Minsky's descendants) won't turn off because it's trying to finish the piece, explaining that the music is a sound sequence evolving toward completion, which will take place in about 150 billion years, roughly the time left for the present universe. The evolving sound sequence of that particular music will end only with the end of change.

Music, because it exists only in time, is the art form that genuinely expresses and evokes the essence of the universe —change. We humans, who love and make music, are part of the moving patterns of the universe, from the field dance of subatomic particles to the flight of the galaxies. And perhaps, just perhaps, the universe is one enormous piece of music.

IV

THE UNIVERSE FROM QUARKS TO THE COSMOS

The
Usefully Small

Smallness is useful in many ways. In our technological civilization we justly praise the tiny microchip or the little fiber-optic tubes that help doctors find out what's wrong with us.

To get even smaller, consider the carbon atom. It comes in slightly different isotopes, some of them radioactive. Carbon 14 is the carbon isotope with a long half-life of about fifty-seven hundred years. It was discovered by Martin David Kamen and in 1947 was put to use by the American chemist Willard Frank Libby.

Cosmic-ray bombardment of the atmosphere converts some of the nitrogen 14 into carbon 14. New carbon 14 is formed as old carbon 14 breaks down radioactively. The resulting equilibrium insures that a very small quantity of carbon 14 always remains in the Earth's atmosphere. A few carbon 14 atoms are present in the carbon dioxide absorbed by plants during photosynthesis, and some of those carbon 14 atoms become part of the plant's tissues. No matter how small a concentration of these radioactive atoms, their presence and concentration can be determined by counting the beta particles they emit.

After a plant dies, it stops absorbing carbon dioxide and therefore carbon 14, and the carbon 14 it has already absorbed will slowly break down in radioactive decay. Scientists can measure the amount of carbon 14 left in a dead plant, and therefore the amount of time elapsed since the plant died. This means that scientists can find the age of anything made from plants, like wooden houses, scrolls of parchment, clothes, paper, and pieces of coal. Small carbon 14 is very useful indeed.

There are even smaller things that we find useful. What would we do without the photon, for instance? Most plants need photons to grow, and animals need to eat either plants or other animals who do eat plants. If Earth's supply of photons from the Sun is markedly diminished, life suffers, and sometimes there's what's known as a Great Dying, when many species—like the dinosaurs—become extinct.

Smaller than the photon, the neutrino seems at first glance to be singularly useless. It's the smallest discrete particle not made of other things and is produced when a proton is converted into a neutron. The neutrino was named by the famous physicist Enrico Fermi and means—in Italian—"little neutral one." Neutrinos really are neutral—they virtually don't interact with matter. When formed in the Sun's core, they leave immediately at the speed of light and can, still at the speed of light, pass through the Earth, and even through us. But the key word is back in the above sentence—*virtually*.

Some neutrinos hit subatomic particles, an occurrence that gives us information about the subatomic particle. Neutrinos are also useful because they show up in the energy given off by supernovas. Detection of neutrinos from the last visible supernova (in the Magellanic Cloud) will teach us more about these spectacular star explosions.

And last but not least, tiny subatomic particles are themselves made of smaller "things" called quarks. So far, scientists have found three, and they have described them using human words like *color* and *flavor*. All very amusing, but in this case, the very, very small is the most important of all, since they are what makes up all matter.

Quarks and
the Last Particle

The current understanding of subatomic physics is that the entire universe is made up, essentially, of two kinds of particles: leptons and quarks. Each is made up of three pairs, and each pair is referred to as a "flavor."

The three flavors of leptons are: (1) the electron and its neutrino; (2) the muon and its neutrino; and (3) the tauon and its neutrino. That's six leptons. Each particle has its mirror image, or antiparticle, so that's twelve leptons altogether. All of these have actually been detected, and physicists have recently come to the conclusion, through rather complicated reasoning, that these three flavors of leptons are all there are. We have them all! (Of the leptons, the electron and its neutrino are the only ones important in our everyday world, but physicists cherish all the flavors and consider them all indispensable parts of the universe.)

Physicists consider symmetry very important, so they feel the situation with leptons ought to be reflected in the other kind of particles, the quarks. There should be three flavors of quarks, therefore: (1) the up-quark and the down-quark; (2) the strange-quark and the charmed-quark; and (3) the top-quark and the bottom-quark. Each has its mirror-image antiparticle, so there are twelve quarks altogether.

The quarks have never been detected as isolated particles and probably never will be. They cling together very strongly in twos and threes and cannot be pulled apart.

When they cling together, however, they form other, more complicated particles (over a hundred of them are known), and from these larger particles the properties of the constituent quarks can be worked out. It is the first flavor, the up-quark and the down-quark, that are the only ones important in the everyday world, for out of them the protons and neutrons that are found in atomic nuclei are built up. However, physicists, here, too, cherish all the flavors.

From the particles containing the quarks, it is possible to deduce the mass (or heaviness, so to speak) of each individual quark, and, on the whole, the quarks are more massive than leptons. This is important, because mass is a form of energy, and the more massive a particle, the more energy it takes to create it in modern atom-smashing machines, and the more difficult it is to detect.

The up-quark, which is the least massive, has a mass about five times as great as that of the electron, and the down-quark has one about seven times as great. These are low masses, and the up- and down-quarks have been known since the existence of quarks was first pointed out in 1963 by Murray Gell-Mann, who got a Nobel Prize for it.

In the second flavor, the strange-quark has a mass about 150 times that of the electron, and that isn't too bad either, and it has been known since 1963. The charmed-quark, however, is about 1,500 times as massive as an electron, and any particle containing it must be more massive still and so is very hard to form. Particles containing charmed-quarks weren't discovered till 1974. Burton Richter and Samuel Ting, who accomplished the task, got Nobel Prizes for it.

That leaves the third flavor. The bottom-quark has a mass about 5,000 times that of the electron, and it was discovered in 1978.

That leaves the top-quark. Of all the leptons and quarks, it is the last particle, the only one not actually detected. Physicists are convinced that it exists, but they would like to discover or create a particle that actually contains it.

The trouble is that they suspect the top-quark to be at least 45,000 times as massive as an electron, and that amount of mass requires the maximum amount of energy that the world's largest atom-smashers can produce.

It *can* be done. There are certain particles, called W-particles, that are not actually constituents of matter but that are needed to make other particles interact in certain ways. These W-particles are roughly as massive as the top-quark, and yet they were detected in 1983 by Carlo Rubbia, who got a Nobel Prize out of it.

The W-particles were searched for both by Fermilab in Illinois and by the CERN atom-smasher in Geneva, Switzerland, and it was CERN that won that race. Fermilab is now trying to smash particles at higher levels of energy than previously used, hoping to discover particles containing top-quarks among the debris.

If they can't detect it, physicists will get nervous about the entire structure they have so painstakingly built up.

But is the top-quark really the *last* one? Actually, probably not. Associated with the W-particles I just mentioned is the "Higgs particle," about which very little is known. It has not been detected, so far. And, of course, there may be other particles that physicists, so far, only speculate about.

The Finding
of the Quark

The 1990 Nobel Prize for physics went to three physicists, Jerome Friedman and Henry Kendall of the United States and

Richard Taylor of Canada, for work that was done twenty years ago.

What happened was this. Ever since 1930, it was known that atomic nuclei consist of protons and neutrons, which were two examples of a type of subatomic particle called a "hadron." The trouble was that scientists kept finding additional hadrons, until over a hundred of them were known, each one different from all the others.

This was a very puzzling phenomenon, and in 1964, the American physicist Murray Gell-Mann worked out a theory in which all the hadrons consisted of a few different, still more fundamental particles that he called "quarks." (Others worked out the same theory independently.) These quarks were combined two or three at a time, and each different combination was a different hadron.

This replaced the chaos of all those hadrons with a kind of sensible and much-simplified order, so that Gell-Mann received a Nobel Prize in 1969.

The problem was that no one could isolate individual quarks or detect them. Many scientists felt that quarks were just a mathematical device to account for the hadrons but that they had no real existence. (Thus we know that a dollar bill can be exchanged for ten dimes. However, if you tear up a dollar bill you won't find any dimes in it.)

How does one find out what the internal structure of a proton (the most common and familiar hadron) really is?

Back in 1911, the British physicist Ernest Rutherford faced a similar problem in determining the internal structure of the atom. What he did was to take a thin film of gold and bombard it with energetic alpha particles from radioactive materials. The alpha particles passed right through the gold film and clouded a photographic film behind it. Since almost all the alpha particles passed through untouched, Rutherford decided the atom was mostly empty space.

However, one out of millions of the alpha particles bounced and was deflected into a new direction. This meant that somewhere in the atom there was something heavy. What's more, because so few alpha particles were affected, that heavy portion of the atom must be exceedingly small and hard to hit. It was from this experiment that Rutherford worked out the fact that at the center of the atom was a heavy "atomic nucleus" only 1/100,000th as large as the atom itself.

It was this nucleus that was made up of still smaller particles, the protons and neutrons, and it was these still smaller particles that might be made up of the even smaller quarks.

As the years passed, physicists developed atomic artillery much more powerful than alpha particles. They had devices that could accelerate subatomic particles to nearly the speed of light. These speeding particles would have enormous energies, and when they struck other particles, much could be discovered from the details of the collision.

At Stanford University in California there is a large "linear accelerator." This is an evacuated tube nearly two miles long. Through it, electrons can be made to hurry. Surrounding magnets keep pushing the electrons along faster and faster and faster until they emerge at the end of the tube with 20 billion electron-volts of energy. (This is a great deal of energy.) Friedman, Kendall, and Taylor began working with this linear accelerator in 1967.

They hurled the energetic electrons against liquid hydrogen, whose atomic nuclei contain only one proton apiece. The experiment thus forced the electrons to strike protons. The electrons were energetic enough to force their way into and through the protons.

If the protons were made up of "proton-stuff" evenly distributed through the particle, the electrons wouldn't be deflected much. If, however, the protons were made up of quarks, then an electron that hit a quark might be deflected quite a bit—as in Rutherford's experiment but on a much more energetic scale.

By 1968, Friedman and the others were detecting the type of deflection that made it seem there were particles inside the proton. For years, they continued their experiments in an attempt to determine what the characteristics of those particles were. Other scientists, notably Richard Feynman, joined in this attempt.

By 1974, it was clear that the particles inside the proton were just like the particles that Gell-Mann had proposed. Some of the properties were very strange—like the fact that quarks were the only particles known to have a fractional electric charge, something that most scientists would have said was impossible.

Now we know a great deal about quarks and about their properties. We know how they combine to form hadrons, and, in short, we know much more about the universe than we did.

Quark Globs

Ordinary matter is not very dense. Water, for instance, has a density of one gram per cubic centimeter. That is because the really dense part of the atom, the proton at the center, is kept away from other protons by electrons.

Some elements have nuclei consisting of many protons and neutrons all clinging together. These nuclei are kept apart by electrons, but even so such elements are denser than water. The metal osmium has a density of twenty-two grams per cubic centimeter, for instance.

Matter at the center of a star like the Sun is subjected to so much heat and pressure that the atoms are torn apart and the nuclei move around freely, approaching each other much more closely than they can in ordinary matter. Such matter is much denser than anything on Earth and is called "degenerate matter."

When a star explodes, part of it can collapse into a ball of degenerate matter, and it becomes a "white dwarf." In that case, its size is usually smaller than that of Earth, but it contains as much mass as the Sun. All that Sun's mass squeezed into the space taken up by a small planet—you can imagine how dense that matter must be.

Yet that is not the ultimate. Even in a white dwarf, the nuclei are kept apart to some extent by electrons. However, if the white dwarf is large enough and massive enough, the nuclei simply collapse to the point where the electrons cannot hold them back. Then protons are turned into neutrons. The neutrons have no electric charge and do not repel each other. All the neutrons therefore collapse until they are touching, and the result is a "neutron star."

A neutron star has the density of a neutron, which amounts to 15,000,000,000,000,000 grams per cubic centimeter. A neutron star can squeeze the mass of the Sun into a small globe of perhaps fourteen kilometers (eight miles) across. Such neutron stars were discovered in 1969.

But neutrons are not individual particles. They are composed of three quarks each, and there is the possibility that as neutrons are squeezed closer and closer together, they break up into the individual quarks that can squeeze together still more closely and produce a star that is still denser. Such a quark star would be the densest possible material made up of matter. (Even the quarks can smash, and when that happens, the star simply shrinks to nothingness, though it retains its mass. It becomes a "black hole.")

We can detect neutron stars because they tend to give off tiny bursts of radio waves as they rotate very rapidly. There are some neutron stars that rotate so rapidly that they send out bursts every few thousandths of a second. At such a rate of turning, even a neutron star, with all its density, can hardly hold itself together.

Norwegian physicists T. Overgard and E. Ostgaard believe that if they can find a neutron star that rotates in less than 1/2,000th of a second, it will simply not be a neutron star, but rather a quark star.

Brian McCusker of the University of Sydney in Australia thinks that if quark stars do exist, the quarks would be stable.

No one has ever really detected a quark on Earth, and there are some scientists who believe it is impossible to detect them. On the other hand, it may be that once a quark star forms, bits of it will tear loose as it turns. The result would be "quark globs," each made up of perhaps hundreds of quarks.

Such quark globs may be wandering through the universe in considerable numbers, and some may occasionally fall to Earth. Once a glob hits Earth's atmosphere, it may break up into triplets, each triplet forming a neutron or a proton. On the other hand, there might be individual quarks left over, and they would show up in cosmic rays. Individual quarks would have fractional electric charges, something no other particles have, and would be detected in that fashion. There have been several occasions in which particles with fractional electric charges have been reported, but none of these has been confirmed.

So the "hunting of the quark" continues.

We might ask ourselves—if we never really detect a quark, how do we know it really exists?

The answer to that is that so many aspects of nuclear physics are explained by supposing that quarks exist and react with each other in certain ways that it becomes almost impossible to deny their existence. Nevertheless, no matter how sensible it is to suppose they exist, physicists would like to detect one.

One Atom at a Time

We can now play games with single atoms, taking them one at a time and making them spell words. Recently, two IBM scientists, Donald M. Eigler and Erhard K. Schweizer, spelled out "I B M" in single atoms.

Our knowledge of atoms is quite modern. Some ancient Greek philosophers suggested that all matter consisted of extremely tiny atoms, but they had no evidence for it. It wasn't until 1803 that a British chemist, John Dalton, pointed out that if matter consisted of tiny atoms, that would explain the manner in which elements formed compounds. This was called the "modern atomic theory," and all through the nineteenth century, chemists used atoms to explain everything that happened in the test tube.

However, atoms, if they existed, were so small as to be totally invisible. Even the best nineteenth-century microscope couldn't show them, because light waves, small as they were, were large enough to skip over atoms and not show them.

Some scientists insisted, therefore, that although atoms were a useful concept, they did not necessarily actually exist. In 1905, however, Albert Einstein worked out an equation to show how atoms (if they existed) would bombard tiny particles in solution and cause them to move about randomly. In 1913, a French scientist, Jean B. Perrin, used the equation to calculate how large atoms would have to be to produce the motion as observed. It turned out they were about 1/250,000,000th of an inch across. The fact that small particles in solution do move exactly as though atoms were hitting them convinced everyone at last that atoms did exist.

In 1895, X rays were discovered. They were like light waves but much smaller. X rays were small enough to detect atoms, but they were too powerful for the purpose. They went right through matter instead of being reflected as light is. Also, X rays could not be easily focused.

The electron was discovered in 1896, and in 1923 an American scientist, Arthur H. Compton, showed that it consisted of waves that were about the size of X-ray waves. Electrons, however, could be easily focused, and they could be reflected from matter, too. As a result, "electron microscopes" could be built that were much more powerful than ordinary light microscopes. The first crude electron microscope was devised in 1932 by a German engineer, Ernst Ruska, and he received a Nobel Prize for that in 1986 (fifty-four years later!).

Over the years, electron microscopes were refined and made more and more powerful. The ultimate so far came in 1985 when two IBM scientists, Gerd Binnig and Heinrich Rohrer, invented the "scanning tunneling microscope." They received Nobel Prizes in 1986 also.

This new microscope makes use of a thin tungsten needle that is placed almost in contact with a surface being investigated. A tiny electric current shoots electrons out of the needle, which bounce off the surface and show the position of the atoms that make it up. You see each atom as a tiny featureless sphere.

There is a tiny attraction between the atoms of the surface and the tungsten needle, which is only the width of a few atoms away from the surface. If the needle is carefully manipulated, it can pull a particular atom away from the surface.

This works best with atoms of xenon, which are among the largest atoms and which do not stick together very tightly. Xenon atoms are sprayed onto a nickel surface. Ordinarily, they wouldn't stay there, but if the nickel surface and the xenon atoms are cooled down almost to absolute zero, the xenon atoms have so little energy that they stay put on the surface.

They are then pulled away, one at a time, by the tungsten

needle, moved to a different place on the surface, and the needle is withdrawn. The xenon atom stays in the new place. Another xenon atom is brought in, and still another, each put in a particular place, and, when it is done, thirty-five atoms spell out very clearly I B M.

It's a marvelous example of delicate scientific technique, but is it of any use? Right now, it isn't. In the future, though, it may become possible to manipulate single atoms in such a way as to form substances built up of specific combinations ("molecules") of atoms, substances that cannot be formed by ordinary chemical means.

In addition, it may be that the microchips that make our modern computers and other electronic equipment possible can someday be composed of individual atoms carefully placed. This would mean that the microchips would be more "micro" than ever. We would have small computers that would do far more things in far less time and might even be made to rival the human brain in complexity. Someday!

Measuring
the Electron

To check the fundamental rules that govern the workings of the universe, scientists have to make measurements of very subtle properties and do it as accurately as possible. Physicists spend a lot of time working out ways of making these measurements, and, in recent years, they have been switching to new methods that are more delicate than any in the past.

For instance, consider the electron. It is the most familiar of all the subatomic particles (particles that are far smaller than atoms). Electrons make up part of all atoms and can be easily broken loose from them. The electric current is the result of a flow of electrons, for instance, which is why the particle is called an "electron" in the first place.

According to a theory worked out as long ago as 1930, there ought to be another particle, exactly like the electron, but opposite in electrical charge. The electron carries a negative electric charge, and the new particle should have a positive electric charge of *exactly* the same size. The new particle was discovered in 1932 and was called the "positron" because of its positive electric charge.

The theory, which is a very important one, requires that the electron and positron have *exactly* the same size of charge, and that doesn't mean almost. Simple measurements show that the two particles have sizes that are just about equal, but just about isn't enough. If there is the *slightest* variation in size, it would have to be explained by modifying the theory, and that might give us a still more accurate picture of the universe than now exists.

Scientists must therefore measure the size of the charge in both the electron and the positron and do so in the most delicate possible way in order to see if they are ever so slightly different. The usual way is to smash particles together and make them undergo changes in their nature. From the nature of those changes, the properties can be deduced. That, however, is a very rough way, so scientists are searching for ways of doing it with less force. (It is like doctors who might search for tumors by exploratory surgery, but who would rather use X rays or magnetic resonance, both of which yield information without having to use the knife.)

Scientists have learned, for instance, how to trap a single electron or positron under such conditions as would hold it almost motionless for hours or days at a time. A single almost motionless particle can be studied very closely. It is spinning, which means it is carrying its electric charge in a tiny circle, and that produces a magnetic effect that can be measured. An electron and a positron should both produce exactly the same magnetic effect. In 1989, measurements of trapped particles were reported by Hans G. Dehmelt, of the University of Washington in Seattle, to show that the magnetic effect *was* the same to within a few parts in a trillion. That's not *perfectly* exact (perfect precision can never be obtained), but it is closer to exact than any previous measurement.

How accurate is a few parts in a trillion? If you had two huge boulders, one of which weighed exactly twenty tons, and the other weighed twenty tons plus a millionth of an ounce, that would be an agreement of a few parts in a trillion.

Here is something else. An electron appears to be a fundamental particle; that is, it is not made up of still simpler particles, and it can't bc broken down to still simpler particles. If so, the electron should behave as though it has a zero diameter.

A zero diameter can't be measured, but tests can show that the diameter has to be less than a certain amount. Until now the most delicate tests have shown that electrons can't be more than a trillionth of the width of an atom. It would take at least a trillion electrons, placed side by side, to stretch across an ordinary atom. By testing single electrons that are virtually motionless, scientists now show that electrons have a diameter that can't be more than a thousandth of this. In other words, it would take at least a thousand trillion electrons, side by side, to stretch across an atom.

That's still not zero diameter, of course, but it's closer to zero than we've ever come before and strengthens the support of physicists' theories about subatomic particles.

In general, the latest high-precision tests have supported theories that scientists have been working with, and you might suppose that they would now dust off their hands and say, "Well, that's close enough," and go on to other things.

That, however, they can never do. Every newly precise measurement merely brings another, more distant horizon into view, one that must be striven for. After all, a still more delicate measurement may possibly reveal some hitherto unsuspectedly tiny discrepancy that will give us a deeper, more satisfactory understanding of the universe. In that sense, science can never end its work, and scientists can only feel grateful for that. No one wants the quest for knowledge to end.

Einstein Is Right Again

Einstein's general theory of relativity was recently retested by Gerald Gabrielse and a team at Harvard University. The theory was first announced in 1916, and it was an amazing feat of imagination, for it had virtually no evidence in its favor. It just seemed to Einstein that that must be how the universe worked.

Thus, when scientists performed an experiment in 1919, noting the position of stars near the Sun during a total eclipse, they got certain displacements because the Sun's gravitation pulled at and curved the beam of light. The displacements agreed with what Einstein had predicted.

Einstein was asked, "How would you have felt if the displacements did *not* agree."

Einstein answered, "I would have been sorry for the Lord God, for the theory is *right*."

One of the fundamental bases of the theory is the suggestion that all objects, regardless of their mass, fall at the same speed (if such things as air resistance are disregarded). Thus, a cannonball and a feather drop at the same rate in a vacuum. This is called the "principle of equivalence."

This was first shown in a very crude way by Galileo four centuries ago, and since then, it has been shown to be true by experiments that have been progressively more accurate.

There is just one way in which the principle of equivalence is, as yet, incomplete. Every particle of matter has an "antiparticle." For the electron, the antiparticle is the "positron"; for the proton, the antiparticle is the "antiproton"; and so on. These antiparticles make up "antimatter."

The question is: Does antimatter fall in the same way matter does, or does it respond to gravity in a somewhat different fashion? After antimatter was discovered, Einstein pointed out that if the general theory of relativity was true, then antimatter would have to fall in the same way matter did. This, however, was just another one of Einstein's instincts, for there was no evidence either way. Nor was it very easy to collect evidence.

Gravitation, you see, is by far the weakest of all the forces we know. We're very aware of its existence because we are acquainted with it in connection with huge astronomic bodies. In a body such as Earth, the tiny gravitational forces for each particle add up until in the Earth as a whole, the force is enormous. If we're dealing with single particles, however, or very small groups of them, gravitation is so weak that it can't be measured, and people who work with such particles ignore the gravitational effect altogether.

Since we only get antimatter as single particles or tiny groups, we cannot really measure the effect on them of gravity directly. Scientists are forced to perform indirect experiments.

For instance, if protons and antiprotons can be shown to have precisely the same masses, then they must be pulled at by gravity in equal fashion. How can one tell if they have the same masses—not approximately, but exactly?

This is where Gerald Gabrielse and his team came in. They had both protons and antiprotons whirled about by magnetic fields and measured how many whirls were made in a second. The number of whirls depended on mass, and it turned out that the masses of the two types of particles were the same to within four parts in 100 million.

From this it could be deduced that Einstein was right again and that antimatter reacted to gravity precisely as matter itself did. (Score another for Einstein's instincts.)

Eric G. Adelberger and his team at the University of Washington in Seattle worked a different kind of experiment altogether. They had been interested in making measurements that would indicate the presence or the absence of a "fifth force," like gravitation, but even weaker. No such fifth force has showed up.

However, in carrying out their experiments, they decided that if antimatter fell in a way different from matter, there would be certain effects in their experiment they could detect. Such effects were not detected, and even though Adelberger worked only with matter and not with antimatter, he nevertheless came to the conclusion that the principle of equivalence held for antimatter.

Are scientists satisfied now with the matter of equivalence and with the way in which antimatter falls? Not quite.

The experiments are indirect. The direct measurements involve a response to magnetism, or to the search for the fifth force. What scientists would like to see would be the particles of antimatter actually falling, actually responding to gravity. Then they would be certain. And yet Einstein, in nearly eighty years, has never been proved wrong, and I don't think he will be in this respect, either.

The Exclusion Principle

It is one of the glories of scientific endeavor that any scientific belief, however firmly it may seem to be established, is constantly being tested to see if it is truly universally valid. This has recently been done in connection with something called "the Exclusion Principle" by a group of scientists led by D. Kekez of Yugoslavia, and the principle has survived.

The story begins in 1913, when the Danish physicist Niels Bohr applied the new "quantum theory" to the atom. He showed that within each atom there were electrons that could only take up certain orbits and no others. This accounted very nicely for some of the ways in which atoms absorbed and emitted energy.

Bohr got a Nobel Prize in 1922 for this, but his theory was rather crude to start with, and, as the years passed, it was extended and refined to account for still finer points in the way atoms absorbed and emitted energy.

Finally, in 1925, the Austrian physicist Wolfgang Pauli pointed out that no more than two electrons could be in exactly the same orbit, and those two could only coexist there if they spun in opposite directions. In this way, electrons were absolutely excluded from orbits that already contained a pair of electrons of opposite spin, and Pauli's rule was therefore called the "Exclusion Principle."

The Exclusion Principle turned out to be extraordinarily useful. It explained the properties of all the different atoms in connection with both their chemistry and their energy relationships. It explained the periodic table of the elements (something chemists greatly value) and did so perfectly.

As time continued to pass, the quantum theory was greatly refined, and it became impossible to think of electron orbits in the way we think of planetary orbits about the Sun. Instead, the nature of electrons became a vague melange of waves that could only be described by mathematical relationships. Nevertheless, Pauli's Exclusion Principle continued to hold, not only for electrons but also for protons and for all other particles classified as "fermions."

Of course, it had to hold, because if it didn't, atoms wouldn't have the properties we now know they have; the universe wouldn't be the familiar universe we experience; and, most important of all, we couldn't exist and we wouldn't be around to worry if there was an exclusion principle or not.

Even so, scientists could not see why the Exclusion Principle had to exist. Even if it did exist and supported the theory of the universe, might there not be an occasional instance when an electron, or some other particle subject to the principle, might force its way into an orbit in which it shouldn't be? It might do this so rarely that scientists, not watching for it, would never notice.

For that reason, scientists try to calculate what might happen if the Exclusion Principle were violated. Certain events would occasionally happen in certain "forbidden" ways, giving off radiation that should not be given off. They therefore design careful experiments in which a very occasional bit of radiation might be detected under conditions that would indicate a violation of the Exclusion Principle.

For instance, an electron outside the atomic nucleus might, on very rare occasions, fall into the nucleus and combine with a proton there to produce a neutron. The Exclusion Principle forbids this, but if it happens, a gamma ray is produced that shouldn't be.

Kekez and his group used the "Liquid Scintillation Detector" under Mont Blanc in the Alps. There they had ninety tons of liquid hydrogen that had been studied for six years for what nuclear events might take place. Any evidence of odd bits of radiation arising from violations of the Exclusion Principle would show up.

The final decision came to this: If nuclear events are taking place at random, not more than 1 in 10,000,000,000,-000,000,000,000,000,000,000,000 would violate the Exclusion Principle. That's 1 event in 10 billion trillion trillion.

Mind you, this is an upper limit. What Kekez and his group are saying is that there are no violations with a greater chance of taking place than that. The actual chance of violation might be much less, and might even be zero, so the Exclusion Principle is still safe.

Nevertheless, scientists will continue looking for violations. After all, Kekez and his group have dealt with conditions right here on Earth. Perhaps under more extreme conditions, the Exclusion Principle is more easily violated.

For instance, the Sun produces particles called neutrinos, but it produces far fewer neutrinos than scientists expect it to. They don't know why this should be, but some have suggested that in the extreme conditions of the Sun's center, the Exclusion Principle might not hold very well and that this might lead to a neutrino shortage. This is not likely, in my opinion, but it is obviously something that will have to be checked out, if some way can be figured out for doing so.

Superheavy Elements

Atoms are named according to the number of particles contained in their nuclei. Thus hydrogen, with a single proton in the nucleus, is hydrogen 1. On the other hand, uranium, which is the most complicated atom that exists on Earth, contains 92 protons and 146 neutrons, so it is uranium 238.

In the last half century, scientists have managed to produce, artificially, atoms that are more complicated than those of uranium. In general, though, the more complicated the atom, the more radioactive it is and the shorter its lifetime.

The most complicated atom that has been formed is element 109 (one that has 109 protons and 266 neutrons and that still hasn't got a name). After it is formed, it endures only a few thousandths of a second before it breaks up. Consequently, it gets harder and harder to study these artificial elements, and scientists decided for a time that nothing past 109 was worth fooling with.

However, between 1966 and 1972, a Soviet physicist, Vilen Strutinsky, worked up theories of nuclear structure that made it seem that atoms beyond 109 could exist and could be studied. In fact, element 114 seemed so stable that it might last millions of years.

Naturally, scientists began looking for element 114 at once. They felt that it might have been formed by natural processes and that somewhere on Earth there might be little pockets of it. No such luck. It has not been found.

Then, too, it might be formed artificially as the other heavy elements have been. These elements beyond 109 are often referred to as the "superheavies." The thing to do is to form an element with 114 protons and 184 neutrons. This would be element 298, and according to theory it should be fairly stable.

One way of producing an artificial element is to bombard an existing one with neutrons. A neutron can (once in a long while) enter a nucleus and join with what's there to form a more complicated nucleus. Elements through 100 were formed in this way.

However, the more complicated the atom, the harder it is to insert a neutron into the nucleus. Consequently, these atoms are bombarded with heavier particles such as the nuclei of various small atoms. What's more, these nuclei must be very energetic to work their way into the heavy nuclei, so they are accelerated by means of various atom-smashers. Thus, element 109 was formed by bombardment with oxygen, chromium, and iron nuclei. What's more, it had to be formed one atom at a time.

How to form element 114? Physicists have come up with various theories as to what would be required. The chances of a nucleus moving into a much heavier nucleus and forming element 114 seem very small. It is calculated that less than one event in more than a billion will result in an atom of 114. The remaining reactions produce a fission reaction that destroys the atoms. Perhaps no more than three atoms per day can be produced in this manner.

There have been more than twenty-five attempts reported to synthesize element 114. All have failed. The number of 114s produced may be about three in a million million reactions and this is not large enough for even our best instruments to detect. Furthermore, even if element 114 were isolated, there would be the problem of proving that it was indeed element 114.

The lighter superheavies, say up to element 100, can be identified chemically, but beyond that atomic lifetimes are too short for chemical reactions to be helpful. Scientists have to study radioactive reactions in order to differentiate one superheavy from another. As long as elements decay by producing alpha particles, they are comparatively easy to identify, but element 114 breaks up by spontaneous fission and there are no easy ways of determining that. Scientists, however, are not giving up. A number of them are convinced that the superheavies exist and can be located and studied.

Does it matter? Suppose that element 114 did exist and was discovered and identified. Of what possible use could it be to any of us? Since scientists gladly spend a great deal of money and time in the search, what are they after?

Actually, they have strong ideas as to the structure of the

atomic nucleus. Some nuclei are spherical, some are oval, and the shape dictates their behavior, which is where element 114 comes in. From its behavior, scientists could deduce its shape, possibly leading to a turnover in their ideas about nuclear shape and behavior—especially if the superheavy was quite stable.

This is something scientists are interested in, so they don't ask questions such as, Is it useful? Their question is, What are nuclei like? For that answer, they are willing to invest money, time, and effort.

As Cold as Can Be

A team of French scientists under the leadership of Alain Aspect set a new record for coldness in the spring of 1990. They cooled some atoms of the element cesium to a temperature of 2.5 microkelvins.

What is a microkelvin? It goes back to the British physicist William Thomson, who, later in life, became Baron Kelvin. He pointed out that temperature represented the energy contained by a quantity of matter and that when the energy was reduced to zero, matter became as cold as it could possibly be. It couldn't ever be any colder because you can't have less than zero energy.

This coldest possible temperature is called "absolute zero," and it is set at 273.15 degrees below 0 degrees Celsius, or 459.67 degrees below 0 degrees Fahrenheit. Scientists often find it convenient to measure temperature as so many Celsius degrees above absolute zero.

Thus, ice melts at 0 degrees Celsius (0°C.). We can say it melts at 273.15 degrees Absolute (273.15°A.). However, the Celsius system and the Fahrenheit system of measuring temperature are both named for the scientists who introduced them, so it was felt that the new system should also be named for the scientist. For that reason, scientists say that ice melts at 273.15 degrees Kelvin (273.15°K.). A millionth of such a degree is a "microkelvin."

For many years, scientists have been trying to remove energy from matter to force it to drop to lower and lower temperatures. In 1823, for instance, the British chemist Michael Faraday liquefied chlorine gas at a temperature of 238.7°K.

Scientists continued to suck energy out of matter, usually by liquefying gases and then letting some of it evaporate. That removes energy and lowers the temperature of the unevaporated part. In 1877, oxygen was liquefied at only 90.17°K. Soon afterward, carbon monoxide was liquefied at 81.70°K. and nitrogen at 77.35°K. Then, in 1895, the British scientist James Dewar liquefied hydrogen at 20.38°K.

Things were getting pretty cold, but there was still one gas that defied liquefaction, and that was helium. It was not till 1908 that the Dutch physicist Heike Kamerlingh-Onnes liquefied helium at a frigid 4.21°K. By allowing some of the helium to liquefy, he managed to achieve a temperature of 0.83°K., so that scientists had come within less than a degree of absolute zero.

It was by studying the behavior of matter at liquid helium temperatures that Kamerlingh-Onnes discovered superconductivity, the ability of some materials to conduct electricity without loss. However, the technique of evaporating cold liquids could not produce temperatures less than half a degree above absolute zero. Since a microkelvin is one-millionth of a degree, half a degree is 500,000 microkelvins.

In the 1920s it was found that if certain compounds were

subjected to a magnetic field, all the atoms would line up. If the compound were cooled as low as possible and the magnetic field were then removed, the atoms would fall out of arrangement, and this would consume energy and lower the temperature. Using this technique, scientists got temperatures down to 30,000 microkelvins by 1933.

There are two kinds of helium: helium 3 and helium 4. In the 1960s, methods were worked out to utilize mixtures of the two so as to drop temperatures still further. By 1965, a temperature of only 20 microkelvins (1/50,000th of a degree above absolute zero) was obtained.

How can scientists possibly do better than that? A new technique, involving lasers, was developed in the 1980s. A small group of atoms is subjected to six intense laser beams: up, down, front, back, right, left. The atoms find they cannot move in any direction. There is always the pressure of the laser beam opposing the move. This means that the atoms have no choice but to remain motionless, or nearly motionless. The more nearly motionless they are, the less energy they possess and the lower their temperature.

It is in this way that a temperature of only 2.5 microkelvins was obtained (1/400,000th of a degree above absolute zero).

Can scientists actually *reach* absolute zero? No! One of the basic laws of thermodynamics says they can't. No matter how much energy is abstracted from matter, some is always left behind. In this case, the atoms that are held nearly motionless by laser light can't be held entirely motionless, because they do absorb bits of energy from the lasers and then emit it again. The emission gives them a bit of recoil that represents energy.

However, scientists, even if they can't reach absolute zero, can get closer and closer. There are some who are aiming to attain a temperature that is equivalent to a picokelvin (which is a millionth of a microkelvin, or a trillionth of a degree above absolute zero).

Why? What is the purpose? Well, we can never tell what

new phenomena we'll discover. Besides, scientists, like baseball players, enjoy establishing records.

No Gold!

The latest analyses of the gold in seawater, by Kelly Kennison-Faulkner and John Edmond, geologists at MIT, show it to be present in only 1/1,000th the quantity that had been thought earlier. Too bad, but it's not really a tragedy, as I shall explain.

The rivers of Earth pour across the globe's land surface and reach the ocean, bringing with them dissolved matter of all kinds. Most of the solid materials of the land dissolve only very slightly, but even so traces find their way into the ocean and then remain there.

They don't necessarily remain there forever, for shallow parts of the ocean sometimes dry up, leaving behind what we now call "salt mines," which contain traces of all the various dissolved materials in ocean water in addition to the salt that is the major component. Even allowing for this occasional drying and removal, a little bit of every known element is found in ocean water—including gold.

For thousands of years, human beings have searched the world for gold, and dramatic gold finds on land have resulted in short-lived booms in California in the 1850s, in the Klondike in the 1890s, and so on. But what about the gold in the ocean? Isn't that the biggest mine in the world?

So people have thought in the past, and the ocean mine

has even given rise to dreams of wealth. After World War I, when Germany was required to pay a huge indemnity (which, in fact, she never paid), it occurred to a Nobel Prize-winning chemist, Fritz Haber, that he might develop a method for extracting gold from the ocean and use that to pay off Germany's debt. He tried—but it didn't work.

It turned out that there was less gold in the ocean than Haber had estimated, and it now seems that there is *much* less.

What Kennison-Faulkner and Edmond did to get new and very delicate analyses was to allow their samples of seawater to trickle through a material known as an "ion-exchange resin." This contains certain electrically charged atoms ("ions") that the resin holds loosely and can easily give up. It replaces those ions with others taken from seawater. In this way, all the gold atoms from the sample of seawater are extracted and can later be freed from the resin.

A wild mixture of ions is absorbed by the ion-exchange resin, of course. The mixture of ions is made to stream through a vacuum in a "mass spectrometer." This takes advantage of their electric charge to expose them to a magnetic field that makes them travel in curved paths. The curve for each kind of ion is different, and in this way, the gold ions all land in one particular spot, separated from the rest, so their quantity can be measured.

It turned out that there is an ounce of gold in every 3 billion tons of seawater. It doesn't matter whether the seawater is taken from the Atlantic or the Pacific Ocean; that's all there is.

The Mediterranean Sea is a little different. It is connected with the Atlantic Ocean only by the narrow Strait of Gibraltar, so its waters don't mix well with the ocean. Moreover, it is a warm sea, so much water evaporates and this increases the solid content of what's left. There is three times as much gold in a quantity of Mediterranean water as in the same quantity of ocean water. Maybe other landlocked bodies of water are a little richer in gold, but the quantity is never high anywhere—not nearly as high as was thought.

Of course, we must remember that the ocean is vast. The total amount of seawater on Earth weighs about 1.4 million trillion tons. Even if there is only one ounce of gold for every billion tons, it works out that there are fifteen hundred tons of gold in the ocean altogether, and at the present price of gold, this would come to about 1.725 trillion dollars.

There was a time when such a sum would have caused everyone to talk about "untold wealth," but, unfortunately, the wealth is no longer untold. The national debt of the United States is over the 4-trillion-dollar mark, and it won't be many years at this rate that the money our nation owes to its own citizens and to the rest of the world will be twice the value of all the gold in the oceans.

Then, too, one of the reasons that gold has the dollar value it has is because it is rare. If the fifteen hundred tons of ocean gold should suddenly materialize in the various banks and vaults of the world, its price per ounce would drop precipitously and would pay off far less of the fraction of the national debt than you might suppose.

Finally, what counts is not the total amount in the ocean but how thinly spread out it is, and it is *very* thinly spread out. No known technique (and no technique likely to be known in the foreseeable future) can pick out that thinly spread-out gold cheaply enough. To pick out a dollar's worth of gold would cost a thousand dollars or more of effort, as Fritz Haber found out after World War I. So the gold will just have to stay where it is, and we must work out more sensible ways to solve the nation's and the world's monetary problems.

Why Is
the Sky Dark?

Perhaps the astrophysicist Paul S. Wesson of the University of Waterloo in Ontario has solved a problem that has been annoying astronomers for nearly two hundred years.

Back in 1826, a German astronomer, Heinrich W. M. Olbers, pointed out the following—if there were an infinite number of stars, then in no matter what direction one looked, one would eventually see a star. The star might be too dim to see all by itself, but with an infinite number, you would see a cloud of light. In fact, the whole sky would be lit up by this cloud of infinite stars, so that it would all shine as brightly and as hotly as the surface of the Sun. Life would be impossible. Yet, for some reason, the sky is black. This is called "Olbers' paradox."

The easiest way of solving Olbers' paradox is to suppose that there is *not* an infinite number of stars in the universe. There are only so many, which in Olbers' time, were estimated at a few hundred million, with blankness beyond.

However, this view weakened with time. By the 1920s, not only was it known that our own galaxy contained 300 billion stars (thousands of times as many as had been thought in Olbers' time), but in addition there were hundreds of billions of other galaxies, all glittering with stars. The number of stars was still not infinite, but there were so many of them that astronomers wondered how we could look past them at the black sky.

In the early twentieth century, it was discovered that many galaxies were filled with dust clouds and that the universe generally contained thin dust layers here and there. This dust was very efficient at absorbing light, and people began to think that that was the answer to the black sky. There was a great deal of starlight, but it was absorbed by the dust.

A little thought showed this couldn't be so. If the dust stopped the light, the dust itself would heat up in the process until it glowed. Instead of having stars lit up all over the sky, you would have stars *and* dust all lit up, and we would still be facing Olbers' paradox.

But then in the 1920s something else turned up. The universe was expanding. The distant galaxies were receding from us and growing ever more distant. This had an effect on the light they were emitting. As they receded, the light they emitted shifted in the direction of low-energy red light.

This meant that a large percentage of the light emitted by the stars might be shifted to such an extent that their energy content grew very low and they were not capable of lighting up the sky. There it was—the sky was black because the universe was expanding. This was so attractive a suggestion that in my books on astronomy, I used that as a way of solving Olbers' paradox.

But it was wrong just the same. Wesson did a great deal of calculation as to how much light was lost as a result of the expansion of the universe, and he found that too little was lost to account for Olbers' paradox. Even though the universe was expanding, it should still light up the sky and make life impossible.

For a while, it seemed as if no explanation were possible, but one turned up. The galaxies have not been in existence forever, and they don't stay in existence forever. There is a period in which they are born and begin to glow and a period in which they die out and cease glowing. There may, in other words, be a nearly infinite number of stars in the universe, but they don't give off light for an infinite length of time, and what light they do give off has not yet had time to fill the universe.

Wesson's calculations show that this must be so and that this is the explanation of Olbers' paradox. The sky is black only because there hasn't been time for it to light up altogether.

We are alive because we are still in the early days of the universe.

What does this mean for the future? As time goes on, will the universe gradually fill with light until, perhaps trillions of years from now, it will have reached the state where life is impossible anywhere within it?

Or else will the light produced by stars gradually fade out as individual stars die, and as new stars form will the total amount of light remain static? Or perhaps as new stars form less frequently than old stars die, will the light of the universe gradually fade until there is so much darkness and so little light and heat that life will be impossible anyway? Or will the universe undergo other changes still? It's this sort of thing that fascinates astronomers.

Starlight and Dust

For the first time in history, human beings are now able to study pure starlight.

This must seem an odd statement to anyone who has ever looked up at the quiet night sky on a clear, dark night, but it is true.

The trouble with seeing the stars is that we are located in the near neighborhood of one particular star, the Sun, whose light is so enormously greater than anything else we can see that it drowns out everything.

Of course, half the time, our spinning Earth whirls us into its own shadow. The Sun sets, night falls, and the stars emerge.

That, however, is not enough. The Moon is likely to be in the sky, and it reflects sunlight to such an extent that its light is far stronger than everything else in the night sky put together.

In addition to the Moon, there are the bright planets—Mercury, Venus, Mars, Jupiter, and Saturn—which also reflect sunlight and, put together, are brighter than all the stars. For that matter, Uranus, Neptune, and Pluto, although very dim or even invisible to the unaided eye, also contribute sunlight.

And yet there are times when the Moon is not visible in the night sky (that's true half the time, actually), and there are also times when the planets all happen to be on the daylight side of Earth. This does not happen very often, but it is possible, on rare occasions, to see the night sky minus the Moon and the planets.

Can we not see pure starlight, then? Unfortunately, no.

There are tens of thousands of asteroids circling the Sun, even greater numbers of meteoroids, and an uncertain number of comets as well. All of them reflect sunlight, and this significantly dilutes the otherwise pure starlight that we would see.

In fact, there are worse sources of interfering light. The solar system is a system of bodies, large and small, and some of the "bodies" are very small indeed. They are, in fact, nothing more than dust particles. To put it bluntly, the solar system is a dusty place.

Where does the dust come from? It probably comes from slowly disintegrating comets and from larger bodies that occasionally collide and shatter into progressively smaller pieces. Some people estimate that there are 10 trillion tons of dust in the solar system. This is continually being swept up by larger bodies, but it continues to form at the rate of about ten tons per second, so that the total amount remains steady.

What harm does this do as far as starlight is concerned? Well, the dust reflects sunlight and produces a constant faint light on even the darkest night. The light is brightest in the plane the planets travel in, through the constellations of the zodiac. The light is therefore called "zodiacal light."

When a telescope points at the night sky, even if the Moon and all the planets are absent, and even if you don't want to count the light of asteroids and comets, some 40 percent of all the light that reaches the photographic plate is zodiacal light.

What can be done about this? Until now, nothing. However, back in 1972 and 1973, two probes were sent out to Jupiter. They were Pioneer 10 and Pioneer 11. They did their job, went flying on past Jupiter, and are now well beyond the orbits of the planets and are still radioing back messages.

The dustiness of the solar system decreases as one moves away from the Sun, and the two Pioneer probes are now so far from the Sun that the zodiacal light is insignificant. With their backs to the Sun, the planets, and the dust, they are now looking out at the stars and they detect *only* the stars and other objects outside the solar system and nothing more (except possibly for the insignificant occasional flicker of light from a distant comet belonging to the solar system).

Gary Toller of NASA's Goddard Space Flight Center in Greenbelt, Maryland, analyzed the messages sent back by the probes and found that 82 percent of the light one sees, as one looks out at the starry universe, is indeed produced by the stars in our own galaxy; almost all of it from stars that are too faint to see with the unaided eye.

Since all stars seem to exist in the midst of dust, that starlight is reflected from millions upon millions of zodiacal lights. Almost all the rest of the light seen is, therefore, not directly from the stars but is starlight that bounces off the dust.

A small fraction of the light, about 0.6 percent of the whole, arrives from sources outside our galaxy, from the myriads of galaxies that lie beyond.

This information does not seem, at first glance, to tell us much, but, combined with other measurements, it has served to locate the Sun more accurately in our galaxy. There is an imaginary plane that cuts our lens-shaped galaxy in half, and the solar system, it now appears, is just about forty light-years above that plane.

The Bending of Light

A team of astronomers from the National Radio Astronomy Observatory, under the lead of Glen Langston, has located a tiny ring of radio waves in the sky. It was the second such ring to be discovered, and it is one of the most unusual phenomena we can expect to see in the sky. It is called a "gravitational lens."

The first hint that such a thing could exist came in 1916, when Albert Einstein devised his general theory of relativity. According to that theory, light rays ought to be bent if they passed near a very massive object, just as they are bent as they pass obliquely into glass.

The difference is that while light is bent enormously by glass (refraction), gravitational pull bends light only exceedingly slightly. Even so, in 1919, during a total eclipse, it was determined that the stars near the eclipsed Sun (and therefore visible, as they would not ordinarily be if the Sun were not eclipsed) seemed slightly out of position because of such bending.

Suppose some distant source of light sent rays past the Sun on all sides, and the light was bent everywhere so that it came to a focus. If that focus was achieved near Earth, we would see the distant source of light as a circle of light all around the Sun.

This, however, is impossible. Light is bent so slightly that it would come to a focus only over a distance of many light-years. This means that the massive object that focuses the light must be many, many light-years from us, and the source of light must be many, many light-years farther away still.

It was not supposed that objects far enough away could be detected at all, so astronomers felt that while the gravitational lens was an interesting theoretical concept, it was not something that could possibly ever be observed.

But then in 1963, quasars were observed. These are very bright and very distant objects. They are billions of light-years away and can be seen by the enormous light they radiate and even more so by their radio waves. The radio waves have all the properties of light waves, and they, too, can be bent.

Suppose, then, that exactly between a distant quasar and ourselves is something massive that can bend the radio waves of the quasar and focus them in Earth's vicinity. We would see the quasar not as the usual little smudge of radiation but as a ring centering around the massive object that is doing the focusing.

Of course, the massive object may not be exactly between us and the quasar. The radio waves may skim by mostly on one side and very little on the other. Instead of a ring of radio waves, we would simply see a distorted quasar—perhaps a double quasar, one on one side, and a smaller one on the other.

On March 9, 1979, a double quasar was found, the two components of which were very close to each other. What's more, the spectra were so identical that it was natural to assume we were really seeing one quasar, which was being distorted by a focusing effect. Sure enough, on close inspection, it turned out that there was a giant cluster of galaxies in front of the quasar, a cluster so distant it could barely be seen.

Seven more cases of distorted galaxies were detected afterward, but it was not till 1987 that a quasar was found with a galactic cluster exactly in front of it so that it formed a ring of radio waves. There was the perfect gravitational lens or, as it is sometimes called, an "Einstein ring." Now a second Einstein ring has been discovered, and the astronomers have been able to tell how far away the quasar must be that gives rise to the radiation. The estimate is that it is 2.8 billion light-years away.

What's more, they have also detected the object that is doing the focusing. It is a large galaxy that is exactly in between the quasar and ourselves and that has a mass equal to about 300 billion times that of our Sun. In other words, it is a galaxy that is as large as, or perhaps somewhat larger than, our own Milky Way galaxy.

This raises a question about "missing mass." This is one of the two great question marks that are now bothering astronomers. (The other is the mechanism by which the galaxies formed in the first place.)

There are reasons to think that not all the mass in the universe is in the form of objects we can see. If we add up the mass of all the stars in all the galaxies, it seems to turn out that there just isn't enough mass to account for gravitational effects. Some astronomers think that the missing mass is as much as one hundred times greater than the mass we can see, but no one knows what the nature of the unseen mass might be.

Other astronomers are adamant that the missing mass does not really exist, and the argument is loud and furious. In connection with the new Einstein ring, the mass of the intervening galaxy seems to be between eight and sixteen times the mass of the stars we can see in it. This is not actual proof, but it is a blow to the side of missing mass.

Tiny
Twisty Light

One of the greatest puzzles in astronomy is the "mystery of the missing mass," and recently, some astronomers reported that that missing mass may be made visible after a fashion.

The missing mass belongs to objects in the universe that can't be seen or otherwise detected but that astronomers are certain are there. They don't know what it is or what it might be, but they are nevertheless certain it is there.

How? Well, astronomers have studied a large number of galaxies. They have studied the stars and other bright objects that they can see within them. From what they see, they can calculate the mass (that is, the total amount of matter) of the galaxy. They can tell that 90 percent of this mass is concentrated in a comparatively small region at the center of the galaxy.

From this, they can further calculate just how the galaxy ought to rotate. The stars near the center should move quickly; the stars farther away more slowly (that's how the planets move in our solar system). The only trouble is that galaxy after galaxy refused to turn in this way. The farther stars move as quickly as the closer ones. The only way we can explain this is to assume there is additional mass in the outskirts of the galaxy, mass that we can't detect.

Can this be clouds upon clouds of small bodies, too small to shine, but each supplying mass? Can it be vast arrays of massive subatomic particles we have never detected and know nothing about as yet? We don't know, but *something* is there.

Nor is it just a matter of turning galaxies. The galaxies are arranged in clusters of all sizes, some clusters containing dozens of galaxies, as the one we belong to does, while others contain hundreds or even thousands of galaxies.

When astronomers note the clusters, they find they can measure the mass of the individual galaxies, and therefore the strength of the gravitational pull they exert on each other. They can also measure the speed at which the individual galaxies move within the cluster. In every case, though, the amount of gravitation that the galaxies exert on each other does not seem to be sufficient to keep the galaxies from drifting apart, considering the speed at which they are moving.

The only way in which they can cling together is to suppose there is more mass, and therefore more gravitational pull, than can be explained by the matter we see. The larger the cluster, the more that extra mass has to be.

A number of astronomers feel that the missing mass may make up 90 percent of all the mass of the universe, and it is in the last degree frustrating that we can't detect it and don't know what it is.

Is there anything the mass can do that might give us a hint of exactly *where* it is? That would be *something*. In theory, there is.

As long ago as 1916, Albert Einstein, making use of his general theory of relativity, predicted that rays of light would bend when they moved past massive objects that subjected them to a gravitational field. The amount of bending would depend on the amount of mass and its nearness to the light, and Einstein calculated out just how it would all work. His prediction was verified in 1919, and many, many times since, in a large variety of ways. Astronomers are all convinced that light bends just noticeably when it passes near a particularly massive body.

Suppose, then, that you study a very distant galaxy, one that is just about as far away as we can make one out. Its light travels toward us over a space of millions of light-years, and, in doing so, it may occasionally pass through a thick cluster of galaxies. The gravitational field of that cluster may then introduce a tiny twist on that faint bit of light from the distant galaxy.

Astronomers at the AT&T Bell Laboratories at Murray Hill, New Jersey, and the National Optical Astronomy Observatory in Tucson, Arizona, headed by J. Anthony Tyson, announced, on January 18, 1990, that they had managed to do just that, by the use of new and advanced "charge-coupled devices" plus specially developed computer programs used to analyze the results.

From the tiny twisty light, they can tell, they think, just where, within the cluster of galaxies, the missing mass might be. Perhaps using this technique the universe can eventually be mapped, so to speak, and the distribution of missing mass determined—more here, less there. This may succeed in giving us hints as to the nature of that mass, and that, in turn, might explain a great deal about the universe that we do not know as yet.

However, we must not jump too eagerly forward. The new technique is just at the very edge of what can be done, and it will have to be checked by others. In fact, some astronomers have already expressed their doubts as to whether the new technique can be entirely trusted.

Nevertheless, such is the scientific frustration over this "mystery of the missing mass" that there is bound to be excitement over even the smallest step toward a possible solution.

Mapping
the Stars

Star maps have a long history. The first important one was prepared by the ancient Greek astronomer Hipparchus, about 130 B.C. He had spotted a new star in the sky (something we call a "nova" nowadays), and he wanted to make sure that further such phenomena would be quickly recognized. To do that, he listed the 850 brightest stars and gave their positions by dividing the sky into lines of latitude and longitude. Any bright star that was observed that was not on the list would be recognized at once as a new star.

About A.D. 150, another astronomer, Ptolemy, incorporated Hipparchus's data into his own book on the subject and added 170 additional stars. Ptolemy's map was used for fourteen centuries. In the 1580s, however, the Danish astronomer Tycho Brahe prepared the first modern star map. He made use of instruments he himself had designed in order to get locations more exact than the ancients had managed. His map included 788 precisely located stars.

A century later, in 1661, a sharp-eyed German astronomer, Johannes Hevelius, prepared a map containing 1,564 stars, one that was even better than Tycho's. However, Tycho had done his work before the telescope had been invented, and Hevelius had refused to use one.

After Hevelius's time, stars were mapped only with the use of the telescope. This made it possible to pinpoint the stars more accurately than the unaided eye could do. With the telescope it was also possible to see and locate stars too dim for the unaided eye to make out.

The first telescopic star map of importance was prepared

by an English astronomer, John Flamsteed. The map was published in 1725, six years after Flamsteed's death, and it listed 3,000 stars.

Between 1859 and 1862, however, making use of far better telescopes, and devoting himself entirely to the task, the German astronomer Friedrich Argelander prepared a giant map of the stars, cataloging no fewer than 457,848 stars in four large volumes, each star carefully listed by latitude and longitude.

No one at that time would have thought it possible to make a bigger and better star map, but Argelander had worked without the benefit of photography, which had not yet been developed to the point of recording starlight.

Once it became possible to photograph the night sky through large telescopes, it was no longer necessary to work out the position of each star as one watched. One simply took photographs and then recorded the position of the various stars at leisure. By the early 1900s it was possible to have star maps include millions of stars.

A 1989 map prepared by the Space Telescope Science Institute catalogs the position and brightness of 18,829,291 objects. Of these about 15 million are stars in our own galaxy. The remaining 3 to 4 million are in other galaxies outside our own.

Human beings with good vision can only see about 6,000 stars by eye alone. These go down to the sixth magnitude in brightness. (The higher the magnitude, the dimmer the star.) The 1989 map contains all the stars and other objects down to the fifteenth magnitude, but our telescopes can make out objects to the twenty-first magnitude, so the new star map does not include all the visible objects by any means.

Future star maps—from new discoveries—will probably still have inaccuracies, for all the stars are moving. They are plunging along huge orbits that carry them about the center of the galaxy, some in fairly circular orbits and some in elongated elliptical orbits. Our Sun is, of course, also moving.

The net result of all this is that the stars we view are

moving across the sky, this way and that, like a swarm of bees. The motions are very slow, but they represent shifts in position that, while tiny, will certainly throw off new observations.

What astronomers are now doing, therefore, is calculating the apparent motions of all the stars so that they can incorporate that information, too, in new star maps.

The
Planet Finders

Two astronomers, Shude Mao and Bohdan Paczynski of Princeton University, worked out a new method of possibly finding planets circling distant stars. Before this, the method was simple, provided the object circling a star was a large one. For instance, the star Sirius has, circling it, a white dwarf star that is as massive as our Sun, and about two-fifths as massive as Sirius itself. The result is that the white dwarf (Sirius B) has a gravitational pull that is quite enormous.

Ordinarily, Sirius would travel in a straight line, but Sirius B pulls it out of that line and sets up a wavy motion. As a result, Sirius B was discovered long before it was seen. The same is true of Procyon B, a white dwarf that circles the star Procyon.

But now suppose that the object circling a star is planetary in nature, say an object the size of Jupiter. An object like Jupiter would have 1/1,000th the mass of the Sun, and its gravitational pull on the star would be insignificant.

Nevertheless, there would be *something* that could be seen. If we choose a star that is relatively close to us, then we could perhaps detect a wavy motion that we wouldn't see otherwise. This would be especially true if the star were a relatively small one and the planet circling it were relatively large.

Because of this, Barnard's star, only 5.9 light-years from us, showed a very faint wavy trace as it moved, and astronomers calculated that a planet the size of Jupiter was circling it. Other stars of the same sort were also studied, and about half a dozen were found to have planets circling them. There was only one catch—the wavy motions were so tiny that there was a good chance that the telescopic views were off—and so they were. It was finally decided that the stars did *not* have planets circling them.

Is there any way of telling that there are planets circling distant stars that do not involve wavy motions?

This is where Mao and Paczynski come in. They point out that every once in a while a large planet might move in front of the star it circles as seen from Earth. As a result, the combined gravity of the star and the planet will distort the light of the distant star in a characteristic way. This is called a "microlens effect." So far, two microlenses have been discovered, but these involve objects in distant galaxies. What we are looking for are microlenses in our own galaxy that can tell us something about the stars near us.

Such a microlens will also happen if a planet moves not only in front of its own star, but in front of a more distant star. The astronomers judge that such a microlens effect would take place a few times each year per million distant stars. The result will be rapid variations in the nature of the light of the star, the variations being between two and a half and ten hours. In order for this to happen, the planet must pass almost directly in front of the distant star, and this should happen in between 5 and 10 percent of microlens events. This means that astronomers would have to pay strict attention for a year in order to detect even a single microlens event.

Apparently, a microlens effect will be more effective if the planet moves in front of a double star. This isn't a bad thing, since most stars are double stars.

The astronomers figure that up to 10 percent of all microlens events will show that the planetary object has moved in front of a double star. The microlens effect should be slower in front of a double star, the light variations lasting between 0.4 and 1.7 days. In addition, it should be possible to tell from the microlens effect just how large the object is that interposes itself between Earth and the star.

Mao and Paczynski are free to admit that the entire task would be a very difficult one. Nevertheless, there are no other methods, apparently, for detecting distant planets.

It matters if we do or do not make such a detection, because for one thing, it would tell us whether the galaxy is full of planets or not. If it is, it may also be full of life, and that would certainly be of importance to us.

What's more it may give us a notion of the size of the planets. If the planets we find are all Jupiter-size, then that would be rather depressing. We want Earth-type planets, although even by the method of Mao and Paczynski, it seems very unlikely that we will be able to find small planets.

And, in any case, if things go well, then we will have managed to locate planets circling other suns and do so for the very first time. That, too, is something we would cheerfully accomplish, since so far we have an absolutely empty universe where planets are concerned. We know only our own planets, those that circle our Sun, and we must be bound to feel lonely under such circumstances.

The North Star Changes

The North Star (usually called "Polaris" by astronomers) is a byword for constancy. It is located very near the point in the sky that is above Earth's north pole, and it is about that point that the stars seem to circle as the Earth rotates. The North Star, being nearly at the hub of that rotation, remains in just about the same spot in the sky every night of the year. This constancy made it very useful for steering ships at night in ancient times, for the North Star always marked the north, as a kind of astronomical compass.

But, in 1989, Nadine Dinshaw, a young astronomer at the University of British Columbia, showed that the North Star is changing in a particularly important way. No, it isn't that the North Star is beginning to move from its place; it is a matter of its brightness.

Most stars shine steadily, but some of them show variations in their light output and are "variable stars." About a hundred years ago it was found that the North Star was a variable star. It didn't vary noticeably to the unaided eye, but careful astronomic measurements showed that at times it was 10 percent brighter than at other times.

From the regular manner in which the light of the North Star grew brighter and dimmer, astronomers could tell it was a particular kind of variable star called a "Cepheid." Cepheids are called that because an example of that type of star was first discovered in the constellation Cepheus.

The reason that Cepheids vary regularly in light production is that they are pulsating regularly. They grow larger, then smaller, then larger, and so on. It turned out that all Cepheids of a particular brightness pulsate with the same period, a discovery made in 1912 by the American astronomer Henrietta Swan Leavitt.

This was a particularly useful discovery. It meant that merely by measuring the length of time a particular Cepheid changed from bright to dim and back to bright again, its *real* brightness could be determined. If this is compared to its *apparent* brightness, as we see it in the sky, we can tell how far away it must be.

By measuring the distance of various Cepheids in our galaxy, astronomers were able to get a correct notion of its size for the first time and to show that it was 100,000 light-years across.

Particularly bright Cepheids can be seen even in other galaxies that are not too far away from us, and the distance of those galaxies can be measured. For instance, the Andromeda galaxy (the nearest large galaxy to ourselves) was discovered to be 2.3 million light-years away by measuring the periods of its Cepheids. The distances of other comparatively nearby galaxies were also determined, and these distances formed the basis for later estimates of the distance of the farthest objects we can see and of the possible age of the universe.

Naturally, astronomers were interested in what made the Cepheid variables tick. Stars shine steadily for a long time at the expense of the hydrogen that undergoes fusion at their core. Eventually, though, when enough of the hydrogen has fused, the core becomes so hot that the star is forced to expand. The outer layers cool and redden as they expand. The star is then a "red giant."

Some stars go through an intermediate state. Before they actually expand to a red giant, they go through a period of pulsation—a kind of hesitation in which they expand a bit, then fall back, then expand a bit, then fall back, and so on. At this stage they are Cepheids.

Eventually, as the hydrogen in the core continues to be consumed, the pulsations die down because the star becomes more determined, so to speak, in its expansion, and moves on its way to red gianthood. Astronomers who analyzed the nature of the pulsations felt that the period during which a star remained a Cepheid was a short one, and this, in turn, meant that if enough Cepheids were observed carefully, sooner or later one would be caught at the end of that stage of its existence and astronomers could watch the pulsations die out. If the astronomical scheme was correct, the pulsations ought to die out rather quickly—in perhaps ten years.

In the early 1980s, astronomers began to notice that the North Star's variations in brightness were becoming less marked. Instruments for making the necessary measurements were improved, and Nadine Dinshaw, after a close study of the star over a period of eight months, during which she took 237 spectra, analyzing the light of the star, was convinced of the fact. From the spectra, it is possible to tell if the star is pulsating, if its surface is first approaching us and then receding from us. It turns out that the pulsations are only a third as forceful now as they were when the variations were first discovered and they seem to be weakening year by year.

Soon the star will have ceased pulsating altogether, after having pulsated steadily for perhaps forty thousand years (a very short time to astronomers). What next? Will the North Star now expand into a giant? Will its light become redder and brighter? It might, but it will still stay in its position, and it will still be the North Star.

The
Unseen Star

It may be that, almost a thousand years ago, a tribe of Native Americans in what is now New Mexico saw a star in the sky that European astronomers unaccountably missed. At least that is something that Ralph Robbins and Russell Westmoreland of the University of Texas now believe, after studying a burial bowl that was uncovered about fifty years ago.

The star in question was a supernova, a star that exploded in a blaze of brilliance in 1054, and, possibly, on July 4 of that year, in a premature celebration of our Independence Day. It blazed out in unexampled brilliance, high in the sky in the constellation Taurus, the bull, so that it was clearly visible all over Europe. It was part of the zodiac, that portion of the sky of particular interest to astrologers so that medieval skywatchers must have been studying the region.

Nor was the star hard to see. It was brighter than any object in the sky, other than the Sun or the Moon. It was two or three times brighter than Venus, the glorious evening star. It was so bright that it was visible *in the daytime* for twenty-three days. It was visible at night for nearly two years before it faded away and disappeared, and at its brightest it actually cast a shadow.

Yet no one in Europe saw it; or at least no European report of its observation has survived to this day, although there is a notation in an Italian manuscript that *might* refer to it.

But if there was no good record of its sighting, how do we know that such a star was ever in the sky?

For one thing, the colossal explosion of that star left a cloud of debris behind that astronomers call the "Crab Nebula."

It is still expanding, and from the rate of its expansion, astronomers can calculate backward and can tell that the explosion began somewhere about 1054.

For another, one of the reasons that there were no European reports on the star may have been that Europe was just beginning to emerge from the Dark Ages at that time, and astronomy on the continent was at a low ebb. Few people were really studying the sky and reporting what they saw.

However, Europe wasn't all there was to the world. At the time, the world leader in technology was China, and for centuries, Chinese astronomers had carefully recorded the time of appearance of any new star and the exact portion of the sky in which it appeared. *They* reported what they called a "guest star" as having appeared in a year that corresponds with our 1054 and in the spot now occupied by the Crab Nebula. (Nor was this the only nova they reported; the astronomical annals of China record about fifty guest stars during ancient and medieval times.) Japanese astronomers also reported on the appearance of the bright star of 1054.

The supernova of 1054 was sufficiently spectacular, however, so that it need not have been reported only by astronomers in nations with advanced technologies. People who were primitive by modern standards had to be quite aware of phenomena in the sky that governed the changing seasons and would have watched for unusual signs.

The study of the astronomical lore of primitives has become important recently; it is called "archeoastronomy" ("ancient astronomy").

Archeoastronomy made headlines in the 1960s in connection with Stonehenge, the impressive circles of giant stones (some of them fallen) in southwestern England. Reconstructions of what it must have looked like when all the stones were in place have led some people to believe that sighting along certain stones located the rising Sun on the day of the summer solstice. Some have even pointed out that the rock arrangements could have served as a Stone Age observatory capable of predicting the times at which lunar eclipses might be expected.

Similarly, places were found in the American continents where sunlight could penetrate a crevice in such a way that on the morning of the summer solstice and no other, it would illuminate the interior in a particular way.

Now there is the burial bowl that Robbins and Westmoreland have been studying. At its center is the figure of a rabbit, which may well signify the Moon, for many Native American tribes imagine that the dark markings on the Moon represent a rabbit (just as in western tradition they represent a man with a thornbush).

Under the rabbit is a dark circle with rays emerging in all directions, and this seems to symbolize a star. When the supernova of 1054 first appeared, the crescent Moon was in the sky nearby, and to Robbins and Westmoreland, it seems that the star near the rabbit signifies that. Moreover, there are twenty-three rays emerging from the star, and that may represent the twenty-three days it was visible in the daylight. Finally, dating procedures indicate the bowl was made between A.D. 1000 and 1070, which puts it at the right time. None of this is entirely persuasive—but it is possible, isn't it?

The Color
of Sirius

The brightest star in the sky is Sirius, and it shines like a white-hot diamond. A star like Sirius does not generally change its color and its brightness, so it is rather surprising that in ancient and medieval times, it is frequently described as "red."

How is it possible for Sirius to be red?

There are, actually, several possibilities. After all, Sirius consists of two stars, and one of them, Sirius B, is a white dwarf. A white dwarf is an ordinary star that expands into a red giant. The red giant then collapses into a white dwarf star. One can suppose, then, that Sirius B expanded into a red giant and that it is this that made Sirius appear red.

This is not, however, considered reasonable. Sirius B is sufficiently far from Sirius itself so that when it turns red, it doesn't affect the whiteness of Sirius. Besides this, when a star becomes a red giant and then collapses, it usually produces a cloud of material that would be visible for thousands of years. Such a cloud is not visible, so that if Sirius B expanded to a red giant and then collapsed to a white dwarf, that would have taken place many thousands of years ago.

A second possibility is that Sirius and Sirius B are both surrounded by a cloud of gas that dims their appearance and makes them appear red. However, such a cloud of gas would also exist for thousands of years, and the fact that it is not present now would alter the entire situation.

Then, there is still another possibility. Sirius was, in ancient Egyptian times, considered to be a special star whose appearance in the sky was of great importance. The priests would watch for its first appearance at the horizon. When it did appear at the horizon, it would be seen through the horizon fog, and it would then appear red. It was therefore viewed as a red star.

Of course, the fact that medieval astronomers spoke of Sirius as red may well be an error. They may be referring to the star Arcturus, which is nearly as bright as Sirius and which is distinctly redder.

The great astronomers of medieval times were the Chinese. They refer to the change in color of stars, including

Sirius, but there the reason is entirely astrological. The Chinese felt the universe to be astrological, to be guided by the stars, and they therefore turned things round. Instead of noting the events of the world and deciding that they were run by the stars, the Chinese invented changes in the heavens and then decided that things on Earth were going in accordance with that. However, there are Chinese books that definitely describe Sirius as white and as possessing no change in that respect.

This is not to say that stars never change their color. There are stars that do. There are stars that are red giants and that pulsate so that they are sometimes larger and redder than they are at other times. The most familiar star of this kind is Betelgeuse, in the constellation of Orion. Another is the star Mira, in the constellation of Cetus.

Here, though, the change is small and is by no means a shift from white to red.

Then, too, there are stars that do not change their color, but do change their brightness. There is a star called Algol in the constellation of Perseus that grows dimmer and then grows brighter in a fixed pattern. Apparently, Algol is a double star, and one of the stars is much larger and dimmer than the other. Periodically, the dim star moves in front of the bright one, and then the brightness of Algol diminishes. After a while, the dim star moves on and the brightness returns. This is an "eclipsing variable," and there are a number of such stars.

There are also stars that grow dimmer and brighter and do not involve eclipses. They are stars that simply pulsate, first growing larger and dimmer, then smaller and brighter. They are called "Cepheid variables." These are particularly important because they can be used to measure the distance of galaxies.

Finally, there are stars that are indeed red. In fact, the great majority of stars in the heavens are red. They are small stars and are of dim brightness. They are so small that they can only work up enough energy to have a surface brightness of two thousand degrees—as compared with the Sun's fifty-seven hundred degrees. These are "red dwarfs," and about three-quarters of all the stars are red dwarfs.

It is not at all likely that red dwarfs can possibly support life. For that you need a Sunlike star, but only 10 percent of the stars in the Galaxy are Sunlike. Sirius is considerably larger and brighter than the Sun is, but it can't support life, either.

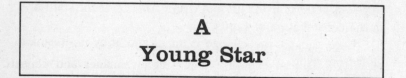

A
Young Star

Colin Aspin of the Joint Astronomy Center in Hawaii and his colleagues have found a star that may be in the process of formation, and may be the youngest star we have yet detected.

It seems odd to think of stars as being of different ages. For centuries, human beings have studied the night sky and have seen stars of all kinds, shining from night to night and from generation to generation without visible change. It would seem they were all created at once in their different brightnesses, but this is not so. With time, astronomers have come to recognize the fact that some stars are small and that some are large; some are cool and some are hot. They shine because they have a supply of hydrogen that gradually fuses to helium.

You might think that the larger the star, and the larger its supply of hydrogen, the longer it lasts, but this is exactly the reverse of the truth. Our Sun is a medium-sized star, and its hydrogen supply would last us for some 10 billion years. About half of it is gone since the Sun is nearly 5 billion years old, but there is plenty left.

A star that is much larger than the Sun contains much more hydrogen but since a large star is very hot, it must use up its hydrogen in great quantities to maintain the heat. The result is that the larger the star, and the more hydrogen it has, the more rapidly that hydrogen is used up, and the star does not last very long.

A large, bright star might last only a couple of billion years, and the largest, brightest stars we know might last only a million years or so. This means that the large, bright stars we see in the sky were not always there, but came into being when the Sun and the Earth were already billions of years old.

If there are stars that only formed a million years ago, why should there not be stars that are forming right now? The answer is, there are, but it is not easy for us to see them.

Stars form out of large volumes of dust and gas. This dust and gas slowly condense and become smaller and denser. Eventually, it becomes so dense at the center that hydrogen fusion starts there. The center "ignites" and becomes a star. In the past, we never saw this happen because the cloud of dust and gas obscured it.

However, nowadays we can study the sky with infrared light and radio waves that can penetrate the dust and gas. The result is that in a cloud of dust and gas called NGC 13333, which is eleven hundred light-years from Earth, little globules of light have been seen. They shine only in the infrared, so they are not as yet true stars, but are "protostars." Astronomers estimate that protostars are only a few thousand years old and that it may be 100,000 years before they condense to the point where hydrogen fusion begins. Of the bits of light in the nebula, one—IRAS-4—is the coolest and therefore is considered the youngest.

Once hydrogen fusion starts, the star gives off a stellar wind, which brushes away all the dust and gas about it. It then shines brightly, and we see it as a star.

There is always a problem as to how large such a star might be. The size of a star depends on the size of the cloud of dust and gas out of which it forms. Such a star can be a small one, dim and cool, shining with red light. (It is called a red dwarf, and most of the stars in the sky are red dwarfs.) Such stars dribble out their hydrogen so slowly that they may last anywhere up to 100 billion years.

Then, of course, there are middle-sized stars like our Sun that are far fewer in number than the red dwarfs. And there are large, giant stars, which last for only a brief period of time and are very few in number compared to the smaller stars that form.

It is possible, though, that there are clouds of dust and gas that are so small that they never condense to the star stage. The center becomes dense enough to form a body that resembles a large planet, but that simply is not large enough to develop nuclear fusion and to become a star. Such "substars" shine by infrared light and are very difficult to see. Astronomers call such bodies "brown dwarfs" because they are not hot enough to shine in red light and become red dwarfs.

Since smaller stars are more numerous, it is possible that there are more brown dwarfs than ordinary stars, and they may add a mass to the galaxy that we don't detect because we haven't been able to spot any brown dwarfs.

Astronomers are busily searching for brown dwarfs because the added mass would answer many problems about our galaxy, but they have not yet been in luck. Every once in a while, a brown dwarf is reported to have been detected but, unfortunately, it turns out to be a false alarm. The search, however, continues and when we spot a very young star, then, for all we know, it may turn out to be a brown dwarf.

Supernovas
I and II

Diana Foss and Richard Wade of the University of Arizona and Richard Green of Kitt Peak National Observatory in Tucson have managed to upset a possible theory for the occurrence of Supernova I.

There are two kinds of supernova—the grand explosions of stars that cause them to glow for a while like a whole galaxy of ordinary stars. One is Supernova I, which is the brighter of the two, and the other, of course, is Supernova II.

Supernova II is the kind we usually think of. It is a large star that is more than eight times the mass of the Sun. Eventually, it runs out of hydrogen to keep it going and it collapses. The collapse sets its outer layers into a wild glow that makes the supernova, and there is usually a small residue that forms at the center—either a neutron star or a black hole.

There is no problem there. The problem is Supernova I, which consists of a star that has already run out of hydrogen and has none left. That means it is a white dwarf. The white dwarf eventually explodes, shining even more brightly than a Supernova II and leaving nothing behind but a cloud of dust and gas. How does a white dwarf explode?

There are plenty of white dwarfs in our galaxy. Fully 10 percent of the stars are white dwarfs. In order for such a star to explode, however, it must be at least 1.4 times as massive as the Sun, and the problem is that all the white dwarfs we know are considerably less massive than that and therefore cannot explode.

The thing to do is to figure out some way in which white dwarfs increase in mass. As they increase in mass, they grow hotter and hotter and more and more unstable, until they reach the 1.4 figure and then they explode. The simplest way is to suppose that there are white dwarfs that exist in pairs and that circle each other rapidly. It is possible that, as they do so, they slowly approach each other until finally the two of them melt together, forming a single white dwarf as massive as the two of them separately. When this happens, the result is a Type I supernova explosion, almost at once.

Foss, Wade, and Green undertook to look for white dwarfs that were sufficiently close to each other and circled each other sufficiently rapidly so that they would eventually combine and produce an explosion. They didn't find a single case of such a phenomenon. The decision then was that Supernova I simply did not result from the collision of two white dwarfs.

Does that leave us with a terrible problem as to how such supernovas form? In my opinion, it doesn't. I don't think there have been any astronomers who really thought that a Supernova I resulted from the collision of two white dwarfs, so that finding that this sort of thing doesn't happen is no surprise.

What can happen? Consider this. There are many white dwarfs that are circling red giant stars. The red giants are so large that their gravitational force does not extend strongly to their very outskirts. The white dwarf, with a much, much stronger gravitational pull (at least in its own immediate vicinity), can therefore pull in some of the material from the red giant.

Consequently, what we have when a white dwarf is circling a red giant is a situation in which material from the red giant is slowly spiraling into the white dwarf. The white dwarf's gravitational force compresses the material it gains and makes it part of its own structure so that with time, the red giant loses mass and the white dwarf gains mass.

Eventually, the white dwarf gains enough mass to surpass the 1.4 limit and it explodes, converting both itself and the red giant it circled into a huge cloud of dust and gas. The red giant contributes mostly hydrogen to the mass, but the white dwarf, having long used up its hydrogen, contributes heavier atoms to the cloud.

This is interesting, for such a "polluted" cloud, one in which there is considerable heavy metal, may eventually collapse to form a central sun, surrounded by planets, which will represent a solar system such as the one we live in.

The Sun and the giant planets are mostly hydrogen, but worlds such as Earth, Mars, Venus, Mercury, and the Moon are for the most part composed of heavier atoms such as those of silicon, iron, magnesium, oxygen, and so on.

So it may be that a Supernova I is not merely a huge sight that we can stare and marvel at. It may be that one of them represented the birth of our solar system, of Earth, and of us. It may be that every element in our surroundings and in our bodies that is not hydrogen was once part of a white dwarf star that eventually exploded, because it circled a red giant star. If so, we have a rather interesting—and violent—history that marks our very beginnings.

Measuring the Distance

Back in 1987, there was a gigantic supernova in the Large Magellanic Cloud. By 1991 that supernova was reported to have yielded a measure of the cloud's distance that is more accurate than any we have ever had before, according to Nino Panaglia of the Space Telescope Science Institute in Baltimore.

Until now, it was estimated that the Large Magellanic Cloud was about 150,000 light-years away, but this was a rather rough figure, and the possibility was that it might be as close as 140,000 light-years, or as far as 180,000 light-years.

As a result of the supernova, which is a gigantic exploding star, a vast amount of dust and gas was hurled out of the star. This dust and gas formed a ring about the star, and as it heated up, it gave off ultraviolet light so that it could be detected.

The ring is tilted at forty-seven degrees to the imaginary plane that connects the remnant of the supernova with the Earth. That means we don't see it broadside as a circle, or edgewise as a line. We see it in between as an ellipse.

Now in order to determine the distance of the star (and therefore of the Large Magellanic Cloud) two things are necessary. First we have to know the apparent diameter of the cloud, and second we have to know its real diameter.

The apparent diameter is easy. It just has to be measured in the telescope. It turns out to be 1.66 seconds of arc. This is not much of an apparent size since the full Moon is something like 1,800 seconds of arc across. In fact, a diameter of 1.66 seconds of arc is something like the separation between two auto headlights viewed at a distance of a hundred miles. Nevertheless, this tiny diameter can be measured quite accurately by astronomers.

But how do you go about determining the real diameter of the ring? This is done by comparing the length of time it takes light to reach Earth from the nearest portion of the ring with the time it takes from the farthest portion.

In studying the records of the supernova, it turned out that light from the near end of the ring first reached us 80 days after the supernova explosion. The light from the far end of the ring didn't reach us till 340 days after the supernova explosion.

Allowing for the tilt of the ring, and the rate at which the ring has been expanding since the days of the supernova explosion, astronomers are able to determine that the real diameter of the ring is 1.37 light-years (nearly 8 trillion miles).

Now astronomers had to ask the question: How far must the ring be so that a real diameter of 1.37 light-years appears to us to represent a width of 1.66 seconds of arc?

The answer comes out 169,000 light-years, which can now be taken as the distance, on the average, of all the billions of stars in the Large Magellanic Cloud. This is rather gratifying, for it comes out fairly close to the crude approximations of the distance made in earlier times.

Nor is the distance of the Large Magellanic Cloud important just for its own sake. For the last sixty years or so, astronomers have been trying to determine the size of the universe, how fast it is expanding, and, therefore, how long ago the Big Bang took place and how old the universe might be. The way to do that is to start with objects that are fairly close to us and then try to estimate from that how far other more distant objects are, and from that the distance of still farther objects, and so on.

The trouble is that in going from one set of objects to another more distant set, astronomers have to make certain assumptions, and they can never be sure how nearly right those assumptions might be. In consequence, the farther out they go from Earth, the more uncertain they are about distances, and about the rate of expansion, and about the age of the universe.

The age of the universe is usually said to be 15 billion years, but that is by no means certain. It could be as little as 10 billion years and as much as 20 billion. In fact, recent studies have shown that galaxies exist in such large clumps that even 20 billion years may not offer enough time for those clumps to have formed. Now, though, with a reasonably exact figure for the distance of the Large Magellanic Cloud, we can work our way outward with at least a better starting point.

That, together with the steadily increasing sophistication of astronomical instruments, may make it possible to get better figures for the distance of far-off galaxies and, by giving us a better figure for the age of the universe, might help us to determine more exactly how galaxies form and clump together.

The supernova of 1987 has thus offered astronomers a remarkable bonus, one that came quite unexpectedly.

Star Clusters

... Look how the floor of heaven
Is thick inlaid with patens of bright gold.
—*The Merchant of Venice*, act 5, scene 1

To the ancients—and that includes Shakespeare—the night sky was unpolluted by man-made light. They could see the stars better than we do, but they did not have our telescopes and so did not know the wonders we can see with technological aid. To the ancients, Earth's sun was the biggest, most magnificent object in the sky. They did not know that our Sun is just another star, moving alone with its planets, or that out beyond the visible stars in the "floor of heaven" there are stars much bigger, more magnificent, and closer together.

The many millions of stars in the galaxies are awe-inspiring, but there are other groups of stars with unique beauty—the clusters, found within a galaxy. After the Cepheid variable stars in clusters were used to determine distance (earlier in this century), there didn't seem to be much else to learn from clusters until now.

In our own Milky Way galaxy, and presumably in most other galaxies, there are thousands of clusters of stars that are held in a group by each other's gravitational influence. The smaller and more common type of cluster is called "open." About one thousand of these are known in our galaxy. They contain from a few dozen to a few thousand stars, loosely arranged. Open clusters are usually located close to the plane of the Milky Way, in or near the galaxy's spiral arms. On the huge time scale of the galaxy itself, open clusters have relatively short life spans, with some stars dropping out of the group as it moves along the galaxy and through dust clouds. The best-known open cluster is the Pleiades, called the "Seven Sisters" by the Greeks, who, despite good eyesight, couldn't see the other three thousand stars in the cluster. It's also possible that our polluted vision (we see now only six with the naked eye) is made worse by the fact that the dimmer one of the seven big stars—Pleione—might have been much brighter in the time of the Greeks.

Much bigger, more closely packed collections of stars are "globular" clusters, some faintly visible to the naked eye as fuzzy patches of luminosity. In the Northern Hemisphere, the brightest seen is M13, the great globular cluster in the constellation of Hercules. In even a small telescope, it shows as a spectacular jewel-burst of up to a million stars. In our galaxy, there are over 125 known globular clusters, moving in a spherical "halo" around the galactic center.

The globular clusters we could examine easily seemed to consist of similar, first-generation stars, with none of the biggest, brightest stars—young blue giants that would have died long before we humans came on the scene. Scientists decided these clusters were almost as old as the galaxy—15 billion years—and unlike open clusters were relatively stable.

Scientists are now rethinking the word *stable*. The large globular cluster 47 Tucanae seems to be producing two rare kinds of stars. One is the "blue straggler," a star that seems to have been given an extra amount of lifetime. The other is the "millisecond pulsar," formed when ordinary stars use up their energy resources and collapse to small dense objects whose radio signals reach Earth as pulses because the star rotates.

Most pulsars rotate about once a second, but these unusual pulsars rotate so fast that the radio beacon is sent at 13 percent of the speed of light. Scientists are now hypothesizing that both the blue stragglers and the millisecond pulsars may be the result of collisions in areas of extreme star density, common in this unusual cluster.

It turns out that not all globular clusters are old. The Hubble Space Telescope has found *young* globular clusters in the elliptical galaxy NGC1275 (near the constellation of Cassiopeia, 200 million light-years from us). The stars of these clusters formed at the same time and are all the same shade of blue, so they can't be more than a few hundred million years old—young on the galactic time scale. Astronomer Jon R. Holtzman, at Arizona's Lowell Observatory, says that NGC1275 may be the product of a collision between two galaxies. The violence of this encounter may have generated the birth of new clusters.

The Hubble telescope has also examined a galaxy named ARP220, which turned out to contain six star clusters that are ten times larger than any in our own Milky Way galaxy. These six are even brighter than the young clusters in NGC1275. Astronomers at the University of Maryland, Edward Shaya and Dan Dowling, postulate that ARP220 was formed by the collision of two spiral galaxies, perhaps 20 million years ago. ARP220 seems to be inherently unstable, with a perhaps dangerously bright and massive center, and with young cluster stars due to suffer supernova explosions.

Well, let's face it. The universe is not a calm and peaceful place. Whatever we human products of the universe achieve in the way of calm and peace is perhaps up to us.

Our
New Neighbor

We have a new neighbor, or, at least, one that has just been discovered. It is a dwarf galaxy in an obscure constellation, called Sextans, in the far south. It was discovered by the use of a telescope in Australia, and its existence was reported in March 1990 by Michael J. Irwin of Cambridge University.

To understand what we mean by a "neighbor," we must begin with our own Milky Way, a dim, luminous band that encircles the sky. This is the "galaxy" (from the Greek word for "milk"), and it consists of a vast assemblage of some 200 billion stars, of which our Sun is one. By far the majority of these stars are hidden behind dust clouds, but astronomers know they are there by their gravitational effects and by means of radio waves reaching us, for radio waves, unlike light waves, can penetrate dust.

When the full extent of the galaxy came to be realized in the 1910s, it seemed sensible to suppose that it constituted the entire universe. After all, 200 billion stars in a system shaped like a pinwheel and 100,000 light-years (600 million billion miles) across is certainly large enough to be a universe.

However, there was a little more than that. Far in the Southern Hemisphere, you can see two foggy regions that look like detached parts of the Milky Way. They are the Large Magellanic Cloud and the Small Magellanic Cloud, and they turn out to be small galaxies that are about 150,000 light-years away, or just outside our galaxy. They seem for all the world to be two satellites of our galaxy, containing only about 20 billion stars apiece.

There were, however, certain small cloudy patches that

could be seen here and there in the sky. These were called "nebulae" (a Latin word meaning "cloud"), and their nature was uncertain. Most astronomers assumed they were dust clouds lying between the stars in our galaxy, but a few heretics thought they might be independent galaxies lying far beyond our own.

In this case, the heretics turned out to be right. The American astronomer E. P. Hubble (after whom the Hubble Space Telescope is named) definitely showed, in 1924, that the Andromeda nebula was a far-distant galaxy. Its distance was eventually found to be 2.3 *million* light-years. It was fifteen times as far away as the Magellanic clouds.

We now know that our mighty Milky Way galaxy is only one of many millions, perhaps billions, of galaxies, strewn through an immensely large space. We have detected galaxies that are *billions* of light-years away from us, so that our perception of the size of the universe has expanded roughly a hundred-thousand-fold in the last three-quarters of a century.

These galaxies do not exist independently but are grouped into "clusters of galaxies." Some of these clusters contain dozens of galaxies, or hundreds, or thousands, or even more.

Naturally, our own galaxy is part of a cluster, a rather small one, that is called the "Local Group." The two best-known members of the Local Group are our own Milky Way galaxy and the Andromeda galaxy. The Andromeda is even larger than our own, containing up to as many as 1 trillion stars.

There is a third galaxy, as large as our own, that was discovered not very long ago. It is called the Maffei galaxy, after its discoverer, and it is located at the very edge of the Local Group. We don't know much about it because it is hidden by dust clouds, but it is about 3.3 million light-years away.

But just as there are many more small planets than large ones, and many more small stars than large ones, there are also many more small galaxies than large ones. The two Magellanic clouds are examples of such small galaxies, which are usually referred to as "dwarf galaxies."

In 1938, the first dwarf galaxy in the Local Group, other than the Magellanic clouds, was discovered in the constellation of "Sculptor." The Sculptor dwarf is about 275,000 light-years away and contains only about 10 million stars. It is very dim for a galaxy and was only discovered because it was so close to us. (That is why we are not as aware of small planets, stars, and galaxies as we ought to be. Only the ones near us can be seen. Objects that are far away can only be seen if they're huge, so we get an unrepresentative sample.)

Since 1938, we have discovered another dozen or so dwarf galaxies that are part of the Local Group. They are each some 3,000 to 20,000 light-years across and are made up of anywhere from 200,000 to 20 billion stars. Even the largest of these dwarf galaxies is only a tenth as large as the real giants.

The new Sextans dwarf galaxy is about 280,000 light-years away, and it is probably one of the smallest. The importance of such dwarfs is this: Astronomers are still puzzling about how the galaxies came to be formed, and since most galaxies are dwarfs, studying them may be more likely to give us answers than studying the rare giants.

Galaxy Update

Galaxy—the word has an aura of glamour, mystery, and adventure. Ad writers, Hollywood, and the science-fiction world have made good use of the word. Didn't one saga take place in a galaxy far, far away?

Real galaxies are just as mysterious. A galaxy is a vast

collection of stars, dust, and gas, held together by gravitational attraction. We live on a planet orbiting a minor star out on one of the limbs of a spiral galaxy we call the "Milky Way." Until the 1920s, nobody realized that there are uncounted billions of galaxies in the universe.

Since the invention of huge telescopes, scientists have been discovering interesting things about galaxies. For instance, most of the galaxies are aging at the same time, since they were born at the time the universe began, but we can discover more about the early years of galaxies by reaching back into space, and therefore into time.

Recently, an "early picture" of a galaxy revealed that it contains a hundred times more gas than our own galaxy, which we see more or less as it is "now." Furthermore, much of the gas is carbon monoxide, which is produced by the first generation of stars. Unless this galaxy is atypical, the finding means that after the Big Bang started the universe going, stars were formed much earlier than we thought.

Quasars, those brightest spots in the universe, are now believed to be powered by supermassive black holes in a galaxy. It seems that black holes are sneaking into all our theories about any kind of galaxy. Here's the latest on elliptical galaxies:

Instead of being smooth, uninteresting aggregations of stars, many elliptical galaxies are peculiar. Some of them rotate, not around a conventional nucleus, but around a different axis, sometimes in the opposite direction. The disorganized aspect of many elliptical galaxies may be the result of the way ellipticals are formed, possibly by the violent merger of two spiral galaxies that contain black holes. It is thought that most elliptical galaxies have massive black holes at their centers.

Then there are starburst galaxies that give birth to new stars more rapidly. Astrophysicists are on their trail, thanks to a new technique for finding supernovas, those spectacular star explosions that produce new stars. The shock waves from a supernova can be tracked by detecting the heated-up dust particles in the path of the wave.

The Milky Way galaxy continues to be exciting. It has a new cousin—another member has been discovered to be part of our particular galactic neighborhood, which consists of our Milky Way, galaxy M31 in Andromeda, and smaller hangers-on—the most obvious to us being the Magellanic clouds. The newly discovered galaxy, named Tucana, is on the opposite side of the Milky Way galaxy from the other members of the Local Group. It's small, faint, and considered to be an elliptical dwarf galaxy. Astronomers are now trying to find out whether Tucana is moving toward or away from the Milky Way.

Last but not, to us at any rate, least—our own Milky Way galaxy. Astronomers are still debating whether or not our own solar system harbors a tenth planet, responsible for oddities in the orbits of Neptune and Uranus. So far the argument is running against the tenth planet, and therefore against a lot of intriguing science fiction plots.

But don't worry, the Milky Way is not a dull place. Astronomers have discovered that our galaxy not only contains a "central exciting source" thought to be a black hole, but coming from the center is a strange bar structure sticking out at a ninety-degree angle. Science fiction writers will probably go full-steam ahead on that one!

Constructing a Universe

There is a problem that is driving astronomers mad, and a graduate student at Princeton University, Changbon Park, has recently tackled it in a dramatic way by using a computer.

The problem is this. By studying the faint radiation of microwaves that comes from every direction in the sky, astronomers are convinced the universe started as a small object that expanded in all directions. The faint radiation is so exactly the same from every direction that it is necessary to suppose that the original small object was homogeneous, that it was perfectly even and had no clumps in it. As it expanded, it perhaps should have remained even and produced a universe that consisted of a smooth ball of gas.

It didn't. It collapsed into clumps that became galaxies, and the galaxies collapsed into clumps that became individual stars. This might have been explained if the galaxies were spread evenly through the universe, but they aren't. They exist in clusters and in clusters of clusters.

Not only that but the galaxies form long lines and curves that surround huge "voids" within which there are practically no galaxies. It is as though the universe were a vast sponge made up of galactic structures surrounding holes.

How could the universe come to take up this shape? What would act to force the originally evenly spread-out matter to take up such an oddly uneven shape?

There are exactly four types of forces in the universe that control all the ways in which objects interact. One or more of them must have shaped the universe. Of the four forces, two —the strong nuclear force and the weak nuclear force—act only within atomic nuclei and have no effect on the universe as a whole. A third, the electromagnetic force, spreads outward for light-years but has the capacity both to attract and to repel, and the two tendencies are practically in balance so that it has little effect on the universe as a whole.

That leaves only the gravitational force to do the work, and astronomers have to figure out a way in which gravitation *alone* can shape the universe. It can be done, but according to calculations it must take time. Some of the larger clusters of galaxies, astronomers calculated, could have been formed by gravity only after a period of time many times that of the actual age of the universe.

Of course, gravitation is probably stronger than we expect. We can only see stars and galaxies, but there is evidence that there is much more gravitational force than that. There is "dark matter," of whose nature we know nothing, that may make up 90 percent or more of all the matter in the universe. However, allowing even for that, it doesn't seem possible that the universe could be created in its present shape by gravity alone—yet there is nothing else to work with. That is the dilemma that astronomers now face.

But working with assumptions and equations on paper may not be enough. It may be useful to attempt a more direct attack—to actually try to create a universe by gravity alone and see what happens. Naturally, this cannot be done in reality. However, the correct programming might allow a computer to *simulate* the construction of a universe. This is what Park of Princeton tried to do, and he presented his results in February 1990.

Park set up a "universe" that was 200 million light-years across. Within this area, he placed 2 million mathematical particles representing clumps of ordinary matter—the stuff out of which stars and galaxies are built. He also added 2 million other particles representing dark matter. He then programmed the computer to allow the particles to move as though they were each being pulled by the gravitational effect of all the others, according to the known rules of how this force works.

It was the largest such simulation ever done, and behold, on the computer screen, the dots of matter formed into structures that looked like galaxies arranged in long curved lines surrounding voids in which there were few or none. In short, the simulation produced something like the universe that actually exists.

Of course, even what Park has done has only served to probe the beginnings of an answer. It will be important to work with a much larger "universe," and instead of millions of particles, it would be nice to work with billions or even trillions. Furthermore, one would have to check the programming and make sure that the assumptions behind the instructions were valid.

However, there are astronomers who are laboring to make larger and, perhaps, more valid simulations, and it may be that, as a result, "universes" can be constructed in greater and greater detail and the results compared with reality.

This has only become possible very recently because of the new supercomputers. The coming of still more powerful computers may help solve still more intractable problems.

Too Clumpy

Roger Clowes of the Royal Observatory, Edinburgh, and Luis Campusano at the University of Chile, Santiago, reported in 1991 the discovery of a band of quasars. This is putting the science of cosmology (the study of the universe as a whole) into a tizzy while cosmologists try to work out what it all means.

The universe began in the Big Bang about 15 billion years ago, and it was a tiny object to begin with, even smaller than a subatomic particle. It must have been very smooth over that tiny volume and it seems that as it expanded, it should have remained smooth and that the universe should be a more or less even expanse of gas.

But that's not the way the universe is. It exists in clumps.

The gas that made up the universe broke up into galaxy-sized chunks, and each chunk then broke up into anywhere from billions to trillions of stars. Therefore, one of the chief jobs facing cosmologists is to account for the method whereby the smooth original universe broke up into the clumpy universe we observe today.

The job was made all the harder when it turned out that galaxies didn't exist by themselves. They weren't spread evenly over the universe. Instead, they existed in still larger clumps called "galactic clusters," some of which included thousands of individual galaxies, and which averaged about 3 million light-years across.

In the last few years, cosmologists have been doing their best to pinpoint the location of millions of galaxies to see just how clumpy they are. In 1989, Margaret Geller and John Huchra of the Harvard-Smithsonian Center for Astrophysics in Boston located a particularly large clump of galaxies that was 550 million light-years in extent. They called it the "Great Wall."

That was bad enough, but ordinary galactic clusters, even the Great Wall, are not extremely far from us, so they were formed when the universe was already quite old and there had been time for the clumps to form.

Now come the quasars. In the first place, quasars are the most distant objects we know. Even the nearest quasar is about 1 billion light-years away, and the farthest ones may be up to 12 billion light-years away or so.

Astronomers have taken it for granted that quasars are scattered more or less evenly across the sky because they are so far away that light reaches us only after billions of years, so that we see the universe as it was when it was quite young. This means there could scarcely have been enough time for the universe to get quite so clumpy.

But Clowes and Campusano detected ten quasars forming a band across the sky. They are at a distance so great that, at the time they formed, the universe may have been only one-third the age it is now. The band is about 650 million light-years across, which makes it 20 percent wider than the Great Wall, and it was formed considerably earlier than the Great Wall was. What's more, further investigation may show that this band of quasars is even larger than now thought. Three more quasars are close enough to be members of the group just possibly. And so far only one-third of that section of the sky has been studied so that the existence of still more quasars in the band is a distinct possibility.

By now, the study of the galaxies, particularly this new finding of the band of quasars, has totally upset all the theories that astronomers have had about the formation of the clumps.

Generally, what astronomers had thought was that there was more matter in the universe than could be seen. There was "dark matter," which did not register on any of our instruments, but which was a source of gravitational attraction. This gravitational attraction might have drawn the universe into clumps.

There are two kinds of dark matter—cold (not much energy) and hot (a lot of energy)—but astronomers don't have the slightest knowledge as to just what this dark matter consists of. There are speculations but no real evidence.

Anyway, with the universe clumping so radically at such an early stage in its development, the dark matter, whether hot or cold, seems to be insufficient to explain it all.

As far as I can see, there are only two ways out. Perhaps the Big Bang took place much longer ago than we think so that there has been more time for the quasars to form the band.

The other possible way out is that the events immediately after the Big Bang did not quite go the way astronomers now think they did, but followed some other procedure that hastened the clumpiness. (If so, we don't know what the other procedure was.)

However, puzzles of this sort are always fascinating, and cosmologists now have a chance to think hard about the whole thing and come up with an array of explanations that may possibly make it easier to understand the universe.

Millisecond Pulsar

Back in 1969, a very peculiar type of star was discovered. It was a neutron star, and while it had the mass of an ordinary star, it was extremely small. It was only about eight miles (fourteen kilometers) across. What's more, it had a spin, a very rapid spin, and it turned at a rate of once per second. Naturally, as such a star grew older, it spun more slowly, and there were some that turned only once every four minutes.

But then in 1987, something even more startling was discovered—a pulsar that spun much more rapidly than an ordinary one. It spun six hundred times in one second. It was turning almost once in a thousandth of a second, and it was therefore called a "millisecond pulsar."

Until recently, astronomers had only located thirteen millisecond pulsars, but in 1991, a group of astronomers was able to locate ten new millisecond pulsars.

The question now is just how the millisecond pulsars manage to spin so rapidly. The most likely explanation is that they are the result of the capture by an ordinary pulsar. Ordinarily, a millisecond pulsar loses spin, and in about a million years, its spin has become so ordinary that we can't detect it anymore. However, since it is trapped by an ordinary pulsar, it manages to pick up mass, and this increases its spin. It spins "up," in other words.

All these millisecond pulsars were detected in a star cluster called 47 Tucanae. It has a very dense core of stars, and it may be that the cluster began with a star population more massive than had been thought. Very likely, there will be many more millisecond pulsars located in the cluster.

It may also be that the millisecond pulsars may be an example of "dark matter." Such matter does not emit photons, so it cannot be seen, but it gives rise to gravitational forces.

The millisecond pulsars also give birth to differences in radio waves. This is very possibly the result of gravitational interactions with nearby stars. This may give rise to new information concerning the core of 47 Tucanae.

There is also a new theory as to why star clusters of the type of 47 Tucanae are so free of gas, even though sizable amounts are formed through stellar activity. David N. Spergel of Princeton University suggests that strong winds produced by millisecond pulsars continually sweep gas out of the clusters. Just a few dozen pulsars would suffice to keep a cluster gas-free in that case.

Richard N. Manchester of the Australia Telescope National Facility in Eppin, New South Wales, led the group that discovered the new millisecond pulsars. Manchester pointed out that 47 Tucanae was so dense in stars that millisecond pulsars might be found there. He pointed out that the cluster was only 13,000 light-years away from Earth, which is only half the distance of most star clusters. It was more likely that weak pulsars could be detected. Manchester turned out to be correct in both respects.

Astronomers seem quite certain that millisecond pulsars spin up as a result of their association with ordinary pulsars. However, astronomers have also found that the total number of ordinary pulsars is considerably smaller than the number of millisecond pulsars. They believe it is possible that ordinary pulsars have a shorter life than millisecond pulsars do.

The first millisecond pulsar to be discovered was viewed simply as a peculiarity, but now that we have a crowd of millisecond pulsars, we can consider them as important objects. What's more, millisecond pulsars are probably scattered through the denser globular clusters. It would not be surprising if we were to have millisecond pulsars close enough to the solar system for us to study them with respect to Earth itself.

Naturally, millisecond pulsars would not be found really

near the Earth. They would be a few thousand light-years away, but even so that would be close enough for study.

A millisecond pulsar would be spinning as a result of its association with an ordinary pulsar. The millisecond pulsar spin would be sufficiently fast so as to nearly tear the millisecond pulsar apart, and that in itself would be a matter of interest.

Millisecond pulsars are stars that are completely unlike ordinary stars. They are tiny, they spin at enormous speeds, and they barely manage to give off light. Their gravitational pull is so enormous that light does not easily escape.

The only stars that outdo the millisecond pulsars are the black holes.

Luminosity

In 1991, a British-American team of astronomers was measuring red shifts for fourteen hundred galaxies and came across one with a red shift that placed it far enough away from us so as to make it enormously luminous. In fact, it was 300 trillion times as luminous as our Sun and thirty thousand times as luminous as the entire Milky Way.

This new object is 40 percent brighter than other objects found, but it is unusual in another way. Most luminous objects are luminous in the visible light. This new luminous object, however, emits 99 percent of its energy in the infrared. What can possibly be the cause of all this infrared luminosity?

The astronomers who discovered the phenomenon suggest two possible causes: (1) there is a quasar buried at the core of the cloud producing the energy; or (2) there is a burst of star formation within the cloud.

Either way, the result is important. If there is a quasar in the cloud's core, then it can't last long. The quasar can only last about a million years before its radiation pressure blows away the cloud. The result is that if a quasar is the cause of the luminosity, it must have just switched on (astronomically speaking).

On the other hand, if the luminosity is caused by star formation, then it may be that the cloud is a galaxy in the first stage of its ignition. Either way, astronomers are catching something that is very unusual.

The team suspects the quasars to be the safer bet, but the presence of a quasar would broaden certain spectral lines, and this broadening has not taken place, which tends to weaken the quasar hypothesis. Furthermore, the dust cloud is so massive that that tends to support the star-burst hypothesis. The cloud is composed mostly of metals, and its mass may be as great as a billion times that of our Sun. This would fit a galaxy that was rapidly undergoing star formation.

If this is all true, then starbirth is proceeding rapidly. This upsets astronomical theories that hold that galaxies must form slowly, so astronomers are busily engaged in trying to decide just how galaxies form. Some believe they form in "free fall" and come to exist in less than a billion years. Others believe that galaxies form over a longer time, several billion years at least.

To help resolve the issue, astronomers have been studying globular clusters, thought to be the oldest objects in the universe. Two of them seem to have a 3 billion-year difference in their ages, but since their chemical makeups are also different, this may account for their difference in age. Then, too, one of the globular clusters is composed chiefly of blue giants and the other of red giants. Astronomers are not sure why this should be.

Does it matter whether galaxies are formed rapidly or slowly? Yes, because the question arises as to how the universe is formed—quickly or slowly.

Individual galaxies form either way. The galaxy of which we are part was formed many billions of years ago, billions of years before the Sun and the Earth came into being. Then the Sun and the Earth, and the rest of the solar system as well, all formed out of a vast gas cloud which existed only because the galaxy itself had formed and had cleared up its gas. If the galaxy had not formed in this fashion, then it might well be that the Sun and the Earth would never have formed.

So you see, it is important whether galaxies formed quickly or slowly, since that would mean that the Sun and the Earth would either form or not, as the case might be. If the galaxies formed quickly, then here we are; if they formed slowly, then we might not be here.

Miniblack Holes Everywhere?

Physicist A. P. Trofimenko speculated in 1990 that miniblack holes are everywhere, including here on Earth. But what exactly is a miniblack hole?

The "black hole" is becoming more and more familiar to the general public. It is a piece of matter so crushed together, and so dense, that the gravitational pull in its immediate neighborhood is enormous. Anything that falls in can't get out, so it is a "hole." Even light can't escape, so it is a "black hole."

A black hole is formed when a giant star explodes as a supernova. Part of it collapses, sometimes, into a black hole. Such a black hole is usually somewhat more massive than our Sun. It tends to pick up material about it, without giving any back, so it grows in size, especially if it is formed in a region rich in stars.

Scientists have not located any black holes unequivocally, but are convinced that they exist. It is suspected that at the center of star clusters there are black holes thousands of times as massive as our Sun. There may be still larger black holes, millions and even billions of times as massive as our Sun, at the center of galaxies.

In theory, black holes can come in all sorts of sizes, so that there can be very small ones with masses no larger than an asteroid or less. The trouble is that the smaller the mass of an object the more ferociously it must be squeezed together to form a black hole.

Scientists don't know of any process in the universe today that would form a "miniblack hole," one less massive than the Sun, but back in 1971, the British physicist Stephen Hawking suggested that they could have formed in vast numbers at the time of the Big Bang creation of the universe, when conditions were radically different from what they are today.

Hawking also showed that black holes are not permanent but slowly "evaporate." An ordinary black hole evaporates so slowly, however, that it would take a period of time trillions of trillions longer than the age of the universe so far to disappear.

However, the evaporation rate would increase as the mass of the black hole grew smaller. A miniblack hole would evaporate much more rapidly than an ordinary one would, and as it evaporated, it would evaporate faster still, until it became small enough, when it would go in a puff, releasing gamma rays of a characteristic kind. Perhaps some miniblack holes were formed that were so small that they are now—a mere 15 billion years after the Big Bang—exploding. If so, they haven't been caught at it. The characteristic production of the appropriate gamma rays has not been spotted.

However, just happening to catch one at the moment of explosion would seem to be a very chancy thing. The mere fact that we haven't succeeded (and we've only been watching for a brief time, and very casually) doesn't mean they don't exist.

Trofimenko suggests that they do exist and that some exist in the Earth and that such an existence may be used to explain some geological facts.

For instance, suppose there is a miniblack hole at the center of the Earth. It might be sufficiently massive so that it would account for the high density of the Earth as a whole without geologists having to suppose that there is a large nickel-iron core at the center of the planet. The Earth might be all rock instead.

Again, there are "hot spots" in the mantle of the Earth. A hot spot will stay in the same place while the plates of Earth's crust move slowly along above them. Every once in a while, the hot spot gives rise to a volcanic eruption so that one ends with a series of volcanoes, the older of which, having passed beyond the hot spot, are extinct.

The Hawaiian islands have been formed over one of these hot spots, and the volcano on the island of Hawaii itself, the most recently formed of the group, is still active. Could a miniblack hole buried in the mantle serve as the source of a hot spot? Even a tiny miniblack hole with a mass of only about 6 billion tons would produce enough heat as it evaporated to account for a hot spot.

Trofimenko also points out there are regions of higher-than-usual density on the Moon; there are extinct volcanoes on Mars and active volcanoes on Jupiter's satellite Io. Perhaps all such things are the product of miniblack holes.

Is there any way of checking this suggestion? Trofimenko

points out that miniblack holes ought to produce the tiny particles called "neutrinos." We have learned to detect a few neutrinos that come from the Sun, but a miniblack hole would produce a thousand times as many. Perhaps, then, we can test for neutrino production in places where miniblack holes might exist—say on the sides of active volcanoes.

It's an interesting idea, and worth investigating, but I'll be a killjoy and say I think the chances of its being valid are very small.

The Black Hole Tango

The tango is a graceful Argentinian ballroom dance made popular by the likes of Rudolph Valentino. Ordinarily, it takes two to tango. A black hole by itself, at least from the point of view of current human technology, is virtually invisible. Theoretically, a black hole occurs when a massive star collapses past the neutron-star stage, the immense gravitational field preventing matter from escaping what's called a "singularity." The result is called a black "hole" (first named by physicist John A. Wheeler) because anything coming near it would fall in, forever.

If a black hole happens to be part of the cosmic dance of a binary-star system, and its companion is a normal star, then the gravitational field of the black hole draws matter from the normal star. An "accretion disc" of matter forms around the black hole. As the matter from the "accretion disc" falls into the black hole, it gains kinetic energy that's converted into radiation—X rays. Fortunately, we can detect these X rays.

In our own galaxy, an X-ray source in Cygnus was first detected in 1965. By 1971, irregular changes in the X rays were found. These helped astronomers decide that the X-ray source called Cygnus X-1 was probably a black hole, drawing matter from a huge blue star revolving in tandem with it.

Other possible black holes are now found frequently. Two more candidates have been discovered in our own galaxy. In 1975, the binary system A0620-00 was discovered in the constellation Monoceros. Recently, astronomers found that the binary's visible orange star is revolving rapidly (at least 460 kilometers per second), but at such a distance from its dark companion that the latter must have huge mass. The logical explanation is that the dark companion is a black hole.

The latest black hole candidate found in our galaxy is in the binary system V404 Cygni, 5,000 light-years from us. In 1938, a visible star there had gone nova, and in 1989 it produced a burst of X rays discovered by Ginga, the Japanese satellite. Again, the visible star orbits its dark companion rapidly enough to make it likely that the companion is indeed a black hole. The Spanish astronomers who did the work on V404 postulate that it may be a complex system of three objects, the third being a red dwarf. This makes for a complicated dance, and certainly no tango.

A star's dance with a black hole is dangerous, for it's theoretically possible that the star revolving with a black hole may eventually be completely swallowed up by it. Indeed, recent data from Compton Observatory indicate that powerful bursts of X rays originating outside our galaxy may be caused by stars that are squashed, flattened, heated up, and swallowed, giving off X rays as they fall to their doom.

Lately, there's been a great deal written about black holes in the center of galaxies. Our own Milky Way galaxy may contain a black hole, but we're getting more certain about other galaxies. M32 is a dwarf elliptical galaxy only 2.3 million light-years from our own. It has stars packed together near the core, around which they revolve too rapidly for it to be other than a black hole.

The Hubble telescope shows that a superdense concentration of stars is also being pulled toward the center of galaxy M87, in the Virgo cluster, 52 million light-years away. It's thought likely that most elliptical galaxies contain supermassive black holes, the result of the way ellipticals were probably formed—from the collision of two spiral galaxies—a violent dance!

Astronomers now have a technique for estimating the central mass of galaxies by measuring the gravity they generate. The velocity of revolving stars is determined by watching changes in the wavelength of their light. Astronomers John Kormendy and Douglas O. Richstone studied galaxy NGC 3115, 30 million light-years from us. Apparently, its nucleus is not only spinning fast, but drawing in (and increasing the velocity of) the surrounding stars. This may be the biggest black hole yet discovered and is a far cry from the simple tango of one star dancing with one dark companion. Massive black holes at galactic centers dance with—and probably kill—many stars.

Occasionally, it seems as if the tango is being danced by the astronomers themselves, passing ideas back and forth. An Armenian and an American astronomer both postulate that the galactic centers that seem so violently destructive are quite the opposite. They think nuclei of new galaxies are being generated at the cores.

Whatever's going on with those dancing duos, or with those amazing galactic cores that dance with many stars, the astronomers and astrophysicists are hard at work to find the truth. In the meantime, tango, anyone?

What's
at the Center?

A spot in the constellation Sagittarius lies in the direction of the very center of our galaxy. Energy pours out of that spot in vast quantities, and the question is: What is located at that spot that serves as the source of this energy?

At the outskirts of the galaxy, where our Sun is located in a quiet, suburban neighborhood, stars are spread thinly. As one travels toward the center of the galaxy, however, the stars are squeezed closer and closer together. At the actual center, they must be piled virtually one on top of the other.

Consequently, there are astronomers who feel that the powerhouse at the galactic center consists of a dense star cluster, containing perhaps millions of stars, all blazing away. Evidence for this viewpoint was presented in 1990 by a group of British astronomers under the leadership of David A. Allen.

A large telescope investigating the central spot, called IRS 16, detected floods of infrared radiation of a type that would be expected to be produced by such a star cluster. Near IRS 16, there is also a very hot star, and altogether, Allen's group maintains that that would account for the energy detected at the center.

Astronomers who investigate the area with radio-wave detectors are not convinced. There is a source of powerful radio-wave radiation at a spot very near the galactic center, which they call "Sgr A." Their investigations show that the source of radio-wave radiation is an object less than 2 billion miles across. It is considerably smaller than our planetary system. If the center of "Sgr A" were located where our Sun is, the whole system would only spread out to a little beyond the orbit of Uranus.

The radio astronomers argue that there is no way in which millions of stars could be squeezed into such a small volume and still maintain their identity as separate stars. The stars must, instead, fall together to form one huge star with a mass of some 5 million times that of the Sun, perhaps.

Such a huge star would collapse under its own gravity to a tiny volume, and the gravitational intensity in its near vicinity would be so great that nothing could escape from it, not even light. In other words, the radio astronomers are convinced that the powerhouse of the galaxy is not a vast star cluster, but a giant "black hole."

If such a black hole existed, it would attract matter into itself, and this would spiral round and round and deliver the energy that astronomers detect. What's more, the size of this spiraling matter, "the accretion disc," would be just about the size of "Sgr A."

How, then, are we going to decide which of these two suggestions is correct?

We are in a quandary because we can't actually see the galactic center. It is hidden from us by vast clouds of dust and by other stars. Therefore, we must continue to depend on indirect evidence. We must continue to study the infrared radiation and the radio waves with better and more refined instruments, and perhaps that will enable us to come to some decision.

Does it matter? Yes, it does, for astronomers find that one of their great problems is this: How did the galaxies form in the first place? At the start, the matter in the universe was smooth and even—astronomers are quite convinced of that. In some fashion, however, it grew lumpy, and there is no explanation for that. One possible explanation is that black holes formed and that around each black hole as a nucleus, the stars of a galaxy gathered. But for that to happen, there must be a black hole at the center of our galaxy and, for that matter, at the center of every galaxy. It's a problem that fascinates astronomers.

Of course, one way we might find out the answer definitely is to send a probe out to the center of the galaxy. Its instruments could detect what there is to detect, and it could beam the information back to us, or carry it back physically.

In theory, that would be fine, but there is a great big whopping catch. The galactic center is about 30,000 light-years away from us. This means that if we were to send out a probe at the speed of light (the absolute maximum speed allowed in our universe), it would take it 30,000 years to reach the neighborhood of the center. It might then beam information to us, but at 30,000 light-years the beam would take 30,000 years to reach us, and it would become far too attenuated for us to detect in any conceivable manner in the neighborhood of Earth. The probe might actually return with the information, again at the speed of light, but that would take 30,000 years, too.

In short, then, to get direct information about the galactic center, we would have to be willing to wait 60,000 years at the absolute least. If the probe traveled at only one-tenth the speed of light, a far more likely situation, we would have to wait 600,000 years for information. This would seem to be impossible, and we will have to continue sifting indirect evidence.

The Cosmic Soup

We all know what soup is: a rich, random mixture of many ingredients circulating in a fluid. We eat soup, and a good soup is one of the great gifts of the gods.

However, soup has been expanded as a term to include many things other than a good minestrone. For instance, it is possible that when the Earth was young and there was no life on it, the action of ultraviolet rays on the simple compounds in the primordial atmosphere and ocean built up more complicated molecules. Perhaps the ocean became a "soup" of such molecules, out of which the first primitive forms of life appeared. Not everyone agrees with this picture of the origin of life, but it has been a popular version, and we all know what an "ocean soup" is.

The "ocean soup" is not the first of the species. Most scientists agree that the universe is expanding, so if we reversed time and imagined the universe growing younger, it would be contracting. It would slowly get smaller, denser, and hotter. As time moved back toward the Big Bang, the temperature would rise into trillions of degrees, with a small universe in which all the particles had lost their organization. No molecules, no atoms, just primeval particles, mostly quarks and the particles ("gluons") that hold them together.

This is the "cosmic soup," and you might ask what astronomers know about this soup. The answer is, "Nothing much."

How do they go about finding more? One way is to start with the universe as it is today (and we know quite a bit about that) and then work backward, trying to figure out what hap-

334

pens with each increment of temperature. Such a course is, of course, fraught with uncertainties.

Another way is to try to create a submicroscopic piece of cosmic soup by smashing together heavy particles. In pushing together, they may create the cosmic soup.

Peter Levai and Berndt Miller of Duke University decided that if a cosmic soup were in this way created, it should send out particles of many kinds at high velocities in a direction perpendicular to the incoming colliding particles. All the velocities should be the same—at least so the theory states.

Levai and Miller began with very simple particles: protons and antiprotons. They smashed them together at huge velocities, and, behold, particles came off just as the theory predicted. Apparently, they had formed a cosmic soup.

There are other physicists who are still working with heavy atoms, and one would expect that the cosmic soup would then be produced in more powerful fashion so that the results could be more definite and informative.

And suppose all this works, what do we get out of it?

For one thing, it will help us work out a grand unified theory that combines all the forces of nature—electromagnetic, nuclear (strong and weak), and gravitational—under a single set of equations, something physicists have been searching for for years. For another, it may answer other questions as well.

One of the most annoying questions is just how the galaxies formed. The cosmic soup is very even. How, then, could it break up into galaxies and stars?

Neil Turok in 1989 suggested that the cosmic soup in the first second after the Big Bang produced a kind of froth, out of which the galaxies formed. Apparently, Turok's calculations showed that the galaxies that formed would be just the right size.

It is amazing, when you stop to think of it, how something so arcane as the cosmic soup, which only existed in the first moments of the universe, should have such consequences. That is one of the most exciting aspects of science—when investigating something out of sheer curiosity, you may find answers to questions you had never expected to be answered.

There have been numerous examples of this in science history. Anton van Leeuwenhoek, one of the very early microscopists, was the first to detect bacteria. He hadn't the faintest idea of what they were, nor did anyone else at the time. They were just tiny specks that might be alive. In time, Louis Pasteur showed that those tiny specks carried disease and that dealing with them properly could prevent such disease. The result was that the human life span was doubled.

One can never tell where the twisting line of scientific discovery will lead us. And who would want to tell? It would spoil all the fun.

Far-Out Reality

In 1920 if Isaac had come from a spaceship instead of being born, he'd have seen the reality of the universe better than any Earthman. He'd have been amused by that year's debate between astronomers Curtis and Shapley over whether the Andromeda "nebula" was close (part of our Milky Way galaxy) or far away. Most astronomers thought it was close, but in 1923 Hubble used the new hundred-inch Mount Wilson telescope on Andromeda. He saw individual stars, calculated the distance using stars called "cepheid variables," and found that Andromeda was indeed a separate galaxy.

Early telescopes negated humanity's picture of a simple

vault of heaven in which the field of stars, the wandering planets, the Sun and Moon all moved around the Earth, but the Mount Wilson telescope irrevocably changed humanity's idea of the universe. Soon other "nebulosities" besides Andromeda were shown to be far-off galaxies. Our Milky Way galaxy was not alone, and suddenly the universe was much bigger than anyone thought.

Technology goes on improving humanity's view of the reality out there, but as new technological achievements appear every day, the old philosophical questions linger, for reality depends on how you look at it.

Optical telescopes don't see past the dust in our own galaxy; radio telescopes do. Then in 1983 the new Infrared Astronomical Satellite located 500,000 infrared sources. More recently, the Rosat satellite's X-ray telescope found many thousands of X-ray sources over 90 percent of the sky. The data are being examined to find whether the X-ray sources are objects previously unknown or already found by other telescopes. Rosat's current investigation of X-ray-emitting objects may explain much about the birth, evolution, functioning, and death of stars, as well as what goes on in the cold gas of interstellar space and in the hot gas around and between galaxies.

Accurately describing far-out reality depends not just on what you're looking at, but on the angle of view. Since galaxies are tilted every which way to Earth's line of sight, astronomers have a problem correlating the viewing angle with what seems to be a galaxy's brightness and size. In the past, these studies have apparently come to erroneous conclusions about brightness and size because distance was not taken into account.

Then there's the adventure of trying to find the reality that exists at the center of our Milky Way galaxy, shrouded in dust. Lately, astronomers are more certain that a black hole occupies the center. The question is, exactly where? One radio source and one infrared source are the chief candidates, and the correct one—when found—may indicate the real center.

The known reality of the universe gets more interesting

every day as our ability to distinguish dim objects grows. Stars ten thousand times dimmer than our Sun have been found, one fairly close (only eight times farther out than Alpha Centauri). Very dim galaxies found recently contain few elements heavier than helium, which means that their gas is probably still in the state it was in after the Big Bang. Studying dim galaxies may help us understand what the universe was like early on.

The Hubble telescope, for all its flaws, brings home the fantastic "reality" of the universe. The expanding picture of this reality seems overwhelmingly violent. Stars burn, collapse, explode destructively. Some galaxies are colliding or obviously did in the past. Galactic centers seem to contain black holes that swallow even light itself and are surrounded by packed stars and fiery flowing gas.

The far-out reality of the universe is more than a trifle frightening, making humanity feel lonely and vulnerable here on our one living planet. Human eyes are limited—faint images don't activate the color-perceiving cone cells of the retina—so that even with a good telescope, galaxies appear colorless. But all that far-out gray violence is our misperception.

On color film, the universe suddenly acquires beauty not just of pattern but of color. True, astronomers aren't sure whether the color images from film are "real," and wonder what they'd see with naked eyes if they could visit far out. They wonder why so many nebulae all seem to be exactly the same shade of red. But what matters is that the universe is beautiful.

Beauty—of color, pattern, and meaning—is not only in the eye of the beholder, but in the way the eyes are used and augmented by equipment. Perhaps the most amazing beauty of all is the effect of this far-out universe on the human mind.

Index

Plants (*cont.*)
  and medicines, 89, 221–22, 223
  and photons, 248
  and sun, 150
Plasma, of comets, 147
Plastics, 183–84, 224
Platinum, 173
Pleiades constellation, 309
Pleione, 309
Plerocercoid, 187
Plesiosaurs, 43
Pluto, 130, 146, 157, 279
  satellite, 141
  struck by icy planet, 158, 159
Poland, 56, 57
Polaris, 292
Pollination, artificial, 224
Pollution, 108, 230
  of atmosphere, 104–05, 106, 224
  noise, 191–93
  ocean, 103
  See also Garbage
Polyhydroxybutyrate, 224
Polymer, 203, 208
Polysaccharides, 22
Porcupines, 76
Pork tapeworm, 188
Poseidon (god), 133, 134
Positron, 260–61, 263
"Positronic" robot brain, 206
Potassium, 208–09
  -40 dating method, 70
  vapor, 118–19
Predators, 14–15, 20, 28
Primate(s), 20, 62
  handedness of, 78–80
  and music, 242–43

  smallest, 219
  warning cry, 193
Principle of equivalence, 263–64
Procercoid, 187
Procyon B, 289
Prokaryotes, 9, 10
Prometheus, 129
"Promoter" in genetic code, 87
Protectionism, 230
Protein
  and cell division, 202
  and cell membranes, 21
  dating fossils with, 70–72
  defined, 3–5
  oldest, 5–7
  and origin of life, 40–43
  production of, 40–42, 83
Proteus, 134
Proton(s), 263–64
  in atomic nuclei, 268
  in cosmic rays, 154
  in cosmic soup, 334–35
  and exclusion principle, 266
  as hadrons, 252
  made of quarks, 250
  in neutron stars, 256
  particles inside found, 253–54
  -proton fusion, 153
  in white dwarfs, 255
Proto-oncogenes, 84
Protostars, 301
Protozoan, 186
Przewalski's horse, 57–59
Pterogota subclass, 234
Pterosaurs, 30, 43, 60
Ptolemy, 287
Pulsars, 309–10, 322, 323
  millisecond, 309–10, 321–23

 **MERIDIAN**   **℗ PLUME**   (0452)

SCIENTIFIC THOUGHT

☐ **ALBERT EINSTEIN: CREATOR AND REBEL by Banesh Hoffman with the collaboration of Helen Dukas.** On these pages we come to know Albert Einstein, the "backward" child, the academic outcast, the reluctant world celebrity, the exile, the pacifist, the philosopher, the humanitarian, the tragically saddened "father" of the atomic bomb, and above all, the unceasing searcher after scientific truth.   (261937—$11.95)

☐ **WE ARE NOT ALONE** *The Continuing Search for Extraterrestrial Intelligence.* **Revised Edition. by Walter Sullivan.** In this completely updated version of his bestselling classic, the renowned science correspondent for the *New York Times* examines past breakthroughs and recent discoveries, as well as the latest technological advances to objectively sum up strong evidence for the existence of alien civilizations and shows how we might find and communicate with them. "A fascinating book"—*New York Times Book Review*   (272246—$12.95)

☐ **FIRST CONTACT** *The Search for Extraterrestrial Intelligence* **edited by Ben Bova and Byron Preiss.** The world's leading astronomers confront the ultimate question: Are we alone in the universe? This extraordinary book provides a complete scientific account of what we want to know, what we are looking for, how the search is being conducted, and what the chances are of actually finding aliens.   (266459—$12.95)

All prices higher in Canada.

Buy them at your local bookstore or use this convenient coupon for ordering.

PENGUIN USA
P.O. Box 999, Dept. #17109
Bergenfield, New Jersey 07621

Please send me the books I have checked above.
I am enclosing $_____ (please add $2.00 to cover postage and handling).
Send check or money order (no cash or C.O.D.'s) or charge by Mastercard or VISA (with a $15.00 minimum). Prices and numbers are subject to change without notice.

Card # _____ Exp. Date _____

Signature _____

Name _____

Address _____

City _____ State _____ Zip Code _____

For faster service when ordering by credit card call **1-800-253-6476**

Allow a minimum of 4-6 weeks for delivery. This offer is subject to change without notice